Theoretical Chemistry and Computational Modelling

Modern Chemistry is unthinkable without the achievements of Theoretical and Computational Chemistry. As a matter of fact, these disciplines are now a mandatory tool for the molecular sciences and they will undoubtedly mark the new era that lies ahead of us. To this end, in 2005, experts from several European universities joined forces under the coordination of the Universidad Autónoma de Madrid, to launch the *European Masters Course on Theoretical Chemistry and Computational Modeling* (TCCM). The aim of this course is to develop scientists who are able to address a wide range of problems in modern chemical, physical, and biological sciences via a combination of theoretical and computational tools. The book series, *Theoretical Chemistry and Computational Modeling*, has been designed by the editorial board to further facilitate the training and formation of new generations of computational and theoretical chemists.

More information about this series at http://www.springer.com/series/10635

Antonio Laganà · Gregory A. Parker

Chemical Reactions

Basic Theory and Computing

Springer

Antonio Laganà
Dipartimento di Chimica, Biologia e Biotecnologie
Università degli Studi di Perugia
Perugia
Italy

Gregory A. Parker
Homer L. Dodge Department of Physics and Astronomy
University of Oklahoma
Norman, OK
USA

ISSN 2214-4714 ISSN 2214-4722 (electronic)
Theoretical Chemistry and Computational Modelling
ISBN 978-3-319-87299-5 ISBN 978-3-319-62356-6 (eBook)
https://doi.org/10.1007/978-3-319-62356-6

Softcover re-print of the Hardcover 1st edition 2018

Printed on acid-free paper

This Springer imprint is published by Springer Nature
The registered company is Springer International Publishing AG
The registered company address is: Gewerbestrasse 11, 6330 Cham, Switzerland

To Giovanna and Jeanene

Preface

Energy and mass transfers in chemical processes are an intricate land of adventure in which atoms and molecules compete and collaborate on different paths in a way that challenges intellectual abilities when trying to rationalize unexpected outcomes.

This book has been designed to help the students of the European Erasmus Mundus Master in "Theoretical Chemistry and Computational Modeling" (TCCM) to familiarize with both theoretical methods and compute techniques useful to handle the microscopic nature of chemical processes. Because of this, the level of references, pseudocodes, and text has been kept as simple and as general as possible by leveraging on the experience gained by teaching the subject for years at the home University and at the TCCM intensive course. We have also tried to avoid misprints and inaccuracies through repeated cross checks. Despite that, the book might not be free of errors and we ask the readers the favor of letting us know (our emails are given in the front page) about possible improvements.

In the book, the reader is driven to disentangle elementary events out of the kinetics of complex systems in which reactive and nonreactive processes combine and compete in different ways depending on the interactions and momenta of the species involved. Out of such complexity, we gradually single out and deal with the key features (by leveraging preferentially on elementary gas phase processes) of two-, three-, four-, and many-body collisions. Then, complexity is regained to extend the treatment to large systems by introducing some approximations.

The book starts in chapter one by considering the modeling of rate coefficients in terms of the transition state (TS) approach. From the analysis of the weakness of the TS model (useful for a phenomenological systematization of experimental data although useless for predictions), the efficiency of chemical processes is rationalized in terms of collisions of two structureless bodies using classical mechanics (in which atoms are considered as mass points) and simple model interactions (like pure Coulomb attraction and/or repulsion, hard sphere, mixed attraction at long range plus repulsion at short range (Sutherland, Morse, and Lennard-Jones)). Classical mechanics computational machinery, relying on both analytical and numerical procedures tailored to solve related Newton, Hamilton, and Lagrange equations, are analyzed in this chapter by associating a set of trajectories starting from different initial conditions to the fate of the collision process (like the angle of deflection) and working out the value of quantities of experimental relevance (like cross sections and rate coefficients).

The observed failure of the classical mechanics treatment to reproduce some key features of measured data (like the elastic differential cross section in two body collisions) is traced back to the quantum nature of molecular processes and to the related uncertainty principle. This drives the reader in chapter two to the use of quantum techniques for evaluating the properties of both bound and elastically scattered atom–atom systems. Related quantum treatments are then discussed and analytical solutions are first worked out for some prototype cases to the end of guiding the reader to use of special functions. Then, some basic numerical techniques and related pseudocodes useful for integrating the corresponding Schrödinger equation for generic atom–atom interactions, are illustrated and applied in order to compare related results with corresponding classical ones.

At this point, the reader is ready to abandon the constraint that atoms are structureless bodies and deal in chapter three with the electronic structure of atoms and molecules. To move in this direction, we discuss some techniques used for carrying out ab initio calculations of electronic energies and discuss the adoption of both one electron functions and variational principle. Along this line, the electronic structure of polyatomic molecules, molecular orbitals, Hartree–Fock, and self-consistent field (SCF) molecular orbital (MO) models are discussed in some detail. Then, we end up by illustrating post Hartree–Fock configuration interaction, multiconfiguration self-consistent fields, and perturbation methods for the calculation of electronic energies and other molecular properties. To better deal with larger systems, mention is made also to some empirical corrections simplifying the electronic structure calculations for large sets of atoms as well as for a large number of molecular geometries of the same molecule and a large number of molecules. Finally, the techniques used to shape potential energy global and local functional formulations to fit the distinctive features of computed ab initio values are discussed with the specific intention of attributing to related parameters a physical correspondence.

Next, in chapter four, concepts and techniques to be used for carrying out dynamical calculations of reactive systems starting from atom–diatom elementary processes are considered. To this end, the motion of nuclei is disentangled from that of the electrons by introducing the Born–Oppenheimer approximation. Then, for atom–diatom systems, different sets of coordinates are discussed for singling out those better suited for representing the interaction and for integrating dynamics equations. For the latter, different choices are discussed for classical and quantum treatments as well as for time-dependent and time-independent techniques. The integration of dynamics equations allows to figure out the typical features of the atomistic phenomenology of atom–diatom systems such as the effect of a different allocation of energy to the various degrees of freedom in promoting reactivity, the importance of providing an accurate representation of the potential energy surface, the merits and demerits of adopting reduced dimensionality approaches, or dealing quantally with some degrees of freedom (while handling classically the others). Then, the discussion is extended also to the usefulness of singling out the periodic orbits of dynamical systems for rationalizing their reactive behavior (including the categorization of transition state effects) and designing proper statistical treatments for long living processes.

At this point, the road is paved for considering in chapter five systems of higher complexity starting with the four and more atom ones and ending with those for which the atomistic granularity is difficult to manage with sufficient accuracy. The introduction of additional degrees of freedom, in fact, impacts on the structure and, accordingly, on the formulation of the potential. For this reason, the definition of the quantities to be computed, the computational techniques adopted and the observables to be simulated are also reconsidered. The progress made in this direction is strictly related to the evolution of compute platforms and the level of concurrency and distribution achieved. This has led to a radical change of the organization of molecular sciences toward service-oriented procedures, competitive collaboration, data reuse, and openness.

Accordingly, the book is articulated as follows: in the first chapter, we deal with the classical mechanics concepts and their application to the two-body problem; in the second chapter, we deal with the corresponding (two body) quantum mechanical concepts and treatments; in the third chapter, we move toward the description polyelectronic and polyatomic systems, the calculations of related eigenenergies, and the construction of potential energy surfaces connecting the different arrangements of the molecular system; in the fourth chapter, we tackle the problem of describing the atom–diatom reactive systems and properties and illustrate as well the different methods for rationalizing related mechanisms; in the fifth chapter, we move toward more complex (up to many atoms and many molecules) systems and focus on synergistic multiscale competitive collaboration in the context of recent progress made in distributed computing. Eventually, particular importance is also given to the present evolution toward Open Science by referring to a Horizon 2020 funding proposal for establishing a Molecular science European research infrastructure.

Perugia, Italy
Norman, USA

Antonio Laganà
Gregory A. Parker

Acknowledgements

AL thanks his parents (Vincenzo and Giuseppina) for their support to his education even during the difficult times of the post-Second World War of the mid-twentieth century, his wife Giovanna for her loving understanding of his dedication to science and education, his son Leonardo for his commitment to both work and family, his relatives and friends for their love.

AL also thanks the large number of colleagues he worked with (see J. Phys. Chem 120 (27) 4595 (2016)). The book itself is coauthored and is the result of various long-lasting collaborations. Most of the concepts illustrated in it were better understood by him thanks to such collaborations. Particular help for this book was given by Ernesto Garcia and Stefano Crocchianti not only for producing a large fraction of the jointly published numerical results used as illustrative examples in the book but also for adapting them as figures of the book and by Leonardo Belpassi for revising the electronic structure section. Important for the writing of the book was also the continuous interaction with the students of the Theoretical Chemistry and Computational Modeling (TCCM) Erasmus Mundus Master during the classes on "Mechanisms and Dynamics of Reactive Systems."

GAP thanks his two wonderful parents (Byron and Edna) who taught him the importance of an education. He thanks his dedicated wife (Jeanene) and their eight wonderful children (Steven, Michael, Sheryl, Jennifer, Tamara, Marilyn, William, and Christopher) for their love, patience and example.

GAP thanks his Ph.D. advisor, lifelong mentor, and friend Professor Russell T Pack for many fruitful collaborations. GAP also thanks his postdoctoral advisors Professors Aron Kuppermann and John C. Light for their mentorship and guidance.

This book is part of the series that the teachers of the TCCM Master have planned to publish with Springer.

Contents

Chapter 1
From the Phenomenology of Chemical Reactions to the Study of Two-Body Collisions

This chapter guides the reader through the phenomenology of the simplest kinetics of chemical systems to the modeling of the rate coefficients governing their time evolution. From the analysis of the weakness of the transition state (TS) model approach (that is phenomenologically valid but useless for predicting), the rate of chemical processes is rationalized in terms of collisions of two structureless bodies using classical mechanics. In this way, it is possible to follow the space and time evolution of the colliding partners. The machinery of the related classical mechanics equations (Newton, Hamilton, and Lagrange) is explained and the numerical procedures for associating classical trajectories starting from different initial conditions to the fate of the chemical process is given once the interaction is known. Applications to various popular models of the interaction (hard sphere, repulsive Coulomb, attractive–repulsive potentials, like the Lennard–Jones (LJ) and the Morse) are considered for an analytical and numerical solution of the problem.

1.1 From Kinetics to Bimolecular Collisions

1.1.1 The Phenomenological Approach

In order to build a rigorous theoretical and computational ground for the description and understanding of chemical reactions, one has to scale the treatment of the problem of chemical processes down from the macroscopic phenomenological level (that refers to thermodynamics and kinetics treatments) to the microscopic one (that refers to dynamics treatments). The scaling down starts from confining the analysis to gas-phase homogeneous systems in order to more easily relate the parameters characterizing the time evolution of the system to the variation of the intervening species (say X of concentration [X] or partial pressure p_X) because, as is well known, pressure p_X is related to the concentration [X] and the temperature T by the equation $p_X = [X]RT$. The variation of the intervening species is usually quantified in terms

A. Laganà and G. A. Parker (eds.), *Chemical Reactions*, Theoretical Chemistry and Computational Modelling, https://doi.org/10.1007/978-3-319-62356-6_1

of the time t dependence of the reaction rates $v(t)$ as follows:

$$v(t) = \frac{d[X]}{dt} = k(T)[X]^m, \tag{1.1}$$

where $k(T)$ is the temperature-dependent rate coefficient and the power m is the order of reaction with respect to [X], the reactants' concentration.

A further step toward a scaling down to a microscopic (molecular) level the description of the considered reactive process is the formulation of $k(T)$ in terms of quantities depending on the energy E of the system (and whenever appropriate we consider also its partitioning in the various degrees of freedom) like the cross section σ(E), the probability **P**(E), and the scattering **S**(E) matrices. After establishing such relationships and working out the numerical value of the microscopic quantities using appropriate *ab initio* treatments, one can regain the way back to phenomenology by first relating the computed values to the rate coefficients and then evaluating theoretically the measured signal and concentration of the involved species.

As we shall consider in detail later on, of particular importance for that purpose are the product intensities measured in beam-scattering experiments and generated by single-collision events because they refer to quantities that can be computed using rigorous *ab initio* techniques for a large variety of systems. Such direct theory versus experiment comparison paves the way to the understanding of the microscopic foundations of chemical processes and the consequent accurate evaluation of the averaged kinetics and thermodynamics quantities.

For this reason, the starting point of this book is the analysis of the properties of rarefied (very low pressure p) gases in which single-collision processes with no exchange of energy (isolated systems) and no exchange of mass (closed systems) to the exterior can be treated. The related process is usually written as

$$X \rightarrow W \tag{1.2}$$

with the reaction rate $v(t)$ being defined at a given temperature T as

$$v(t) = -\frac{d[X]}{dt} = k(T)[X]^m = \frac{d[W]}{dt}, \tag{1.3}$$

in which the variation of reactants (consumed) has a negative sign while that of the products (generated) has a positive sign.

A more general case is the one in which more than one species participate to the process (the number of participating species is called Molecularity), like, for example, the bimolecular one in which the reactant species are A and B and the product species are C and D (that is $\alpha A + \beta B \rightarrow \gamma C + \delta D$), leading to

$$v(t) = -\frac{d[A]}{\alpha dt} = -\frac{d[B]}{\beta dt} = \frac{d[C]}{\gamma dt} = \frac{d[D]}{\delta dt}. \tag{1.4}$$

The variation of the intervening species corresponding to that of Eq. 1.1 reads now

$$v(t) = -\frac{d[A]}{\alpha dt} = k[A]^m[B]^n, \qquad (1.5)$$

where $m + n$ is the order of the process with m and n being not necessarily integers. In the particular case of m=0, 1, and 2, the rate coefficient takes, respectively, the following analytical forms:

$$k_0(T) = \frac{[A]_o - [A]}{\alpha(t - t_o)} \quad \textit{for } m = 0, \qquad (1.6)$$

where $[A]_o$ is the concentration of A at the initial time t_o and [A] is its concentration at time t,

$$k_1(T) = \frac{\ln[A]_o - ln[A]}{\alpha(t - t_o)} \quad \textit{for } m = 1 \qquad (1.7)$$

and

$$k_2(T) = \frac{1/[A]_o - 1/[A]}{\alpha(t - t_o)} \quad \textit{for } m = 2 \qquad (1.8)$$

as illustrated in the upper row of Fig. 1.1. A more general formulation can be obtained using the dimensionless variables $\eta = [A]/[A]_o$ and $\tau = k(T)[A]_o^{m-1}t$ by plotting

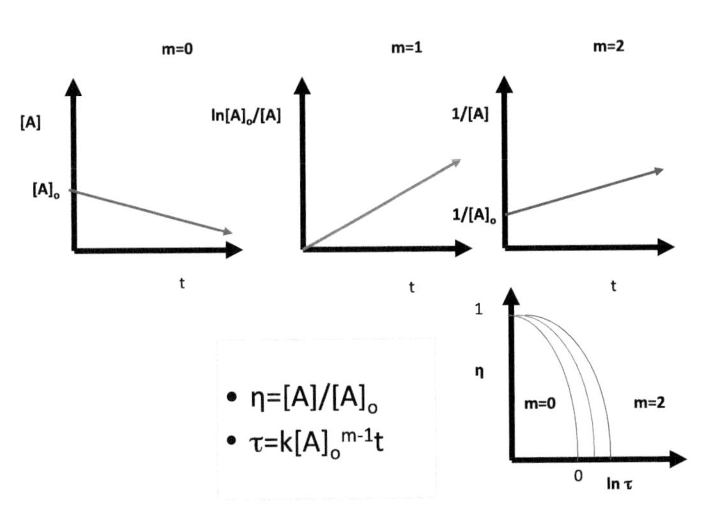

Fig. 1.1 Powell plots for m = 0, 1, and 2 of the concentrations as a function of time t (upper panel) and of η as a function of ln τ (lower panel)

η against $\ln \tau$ (see the example sketched in the lower row of Fig. 1.1 again for $m =$ 0, 1, and 2).

Under the assumptions mentioned above, one can, in principle, estimate the value of the rate coefficient at a different given temperature T and initial concentration $[X]_o$ values (such measurements are performed using either chemical or physical properties of the reactive system) by measuring the current concentration [X] of the involved species at different elapsed times.

1.1.2 Realistic Kinetic Models

However, even seemingly, simple gas-phase reactions are difficult to interpret in this way due to the uncertainty of the experimental measurements and to the complexity of the actual reaction mechanisms. In the real world, chemical processes generally occur through a combination of several different simpler (elementary) steps that produce and connect various intermediates of different stabilities and give rise to complex reaction mechanisms combining initiation, propagation, chain propagation, branching, termination, etc. steps. As an example, let us consider the combustion of pure molecular hydrogen whose mechanism (typically consisting of a set of some tens of elementary chemical reactions) has been reduced, for the sake of simplicity, to those listed in Table 1.1.

In this simplified scheme of the $H_2 + O_2$ combustion process, we can formulate the reaction rate of producing H_2O as follows:

$$v(t) = \frac{d[H_2O]}{dt} = k_2(T)[OH][H_2] \tag{1.9}$$

(hereinafter the dependence of k on T will be dropped when not explicitly required). If we exclude explosion regimes, we can also make the stationary state assumption (equating the rates of production and consumption) for the most important intermediates OH, H, and O and, therefore, we can write

Table 1.1 Reduced set of elementary chemical reactions in which the combustion of molecular hydrogen can be decomposed

Reaction	Role
$H_2 + O_2 \rightarrow 2OH$	(1) Initiation
$H_2 + HO \rightarrow H + H_2O$	(2) Propagation
$H + O_2 \rightarrow OH + O$	(3) Chain branching
$H_2 + O \rightarrow OH + H$	(4) Chain branching
$H + O_2 + M \rightarrow HO_2 + M$	(5) High-pressure termination
$H + \text{wall} \rightarrow \text{H-wall}$	(6) Low-pressure termination

$$\frac{d[OH]}{dt} = 0 = 2k_1[H_2][O_2] - k_2[OH][H_2] + k_3[H][O_2] + k_4[O][H_2] \quad (1.10)$$

$$\frac{d[H]}{dt} = 0 = k_2[OH][H_2] - k_3[H][O_2] + k_4[O][H_2] - k_5[H][O_2][M] - k_6[H] \quad (1.11)$$

$$\frac{d[O]}{dt} = 0 = k_3[H][O_2] - k_4[O][H_2]. \quad (1.12)$$

By summing Eqs. 1.11 and 1.12, one can obtain the concentration of H at the stationary state

$$[H] = \frac{k_2[OH][H_2]}{k_5[O_2][M] + k_6}. \quad (1.13)$$

To the end of eliminating the terms containing [O], we can utilize Eq. 1.12 inside (1.10) and reorder the terms containing the [OH] and [H]. This leads to the expression

$$k_2[OH][H_2] = \frac{2k_I k_5[O_2]^2[H_2][M] + 2k_I k_6[H_2][O_2]}{k_5[O_2][M] + k_6 - 2k_3[O_2]} \quad (1.14)$$

that can be used in Eq. 1.9 to the end of formulating the final rate of the process

$$v(t) = \frac{1}{2}\frac{d[H_2O]}{dt} = \frac{k_I[H_2][O_2](k_5[O_2][M] + k_6)}{k_5[H][O_2][M] + k_6 - 2k_3[O_2]}. \quad (1.15)$$

As mentioned above, in order to obtain Eq. 1.14, it was not only necessary to reduce the considered set of equations and adopt the assumption of the stationary state regime but also to exclude the conditions in the explosion regime.

Accordingly, it makes no sense to use above equations to estimate the values of the k's involved if there is no strict control of the operating conditions. A more theoretically solid approach to the problem of computing the reaction rate is the one integrating in time computationally the system of kinetic equations defining the overall chemical process by possibly taking the full set of intervening reactions (including the less efficient ones especially if they initiate parallel processes generating new species of potential impact on the final fate of the process). This is made possible by the rapid evolution of computer architectures and platforms provided that an, at least theoretical, accurate evaluation of the detailed rate coefficients of the intervening processes can be worked out.

1.1.3 The Transition State Theory Approach

The simplest approach to the evaluation of the rate coefficient of a chemical reaction is the use of thermodynamics data within a transition state theory (TST) treatment.

The thermodynamics treatment assumes the existence of an intermediate ($X^{\ddagger}$) which is in thermal equilibrium with the reactants. The intermediate can then dissociate to form products

$$X \rightleftarrows X^{\ddagger} \rightarrow W. \tag{1.16}$$

The equilibrium constant $K^{\ddagger}$ of such process defined as

$$K^{\ddagger} = \frac{[X^{\ddagger}]}{[X]} \tag{1.17}$$

can be related to k_f via the relationship

$$k_f = \frac{k_B T}{h} K^{\ddagger}. \tag{1.18}$$

This allows us to relate the rate coefficient k_f to the standard free energy change

$$k_f = \frac{k_B T}{h} e^{-\Delta G^{o\ddagger}/RT}, \tag{1.19}$$

thanks to the relationship

$$\Delta G^{o\ddagger} = \Delta H^{o\ddagger} - T \Delta S^{o\ddagger} \tag{1.20}$$

and the Van't Hoff equation

$$\frac{\partial K^{\ddagger}}{\partial T} = \frac{\Delta H^{o\ddagger}}{RT^2} \tag{1.21}$$

with the quantities $\Delta G^{o\ddagger}$, $\Delta H^{o\ddagger}$, and $\Delta S^{o\ddagger}$ being the free energy, the enthalpy (or heat), and the enthropy of activation thermodynamic functions, respectively. By considering that $\Delta H^{o\ddagger}$ is equivalent to the activation energy E_f^o at constant pressure, one can write

$$k_f = \frac{k_B T}{h} e^{-E_f^o/RT}. \tag{1.22}$$

For illustrative purposes, we sketch here (see Fig. 1.2) the simple reactive system of an atom, A, and the diatom, BC, giving the diatom AB and the atom C.

From the basic assumptions of the TST approach, one has the following:

1. The reactive process occurs on a single potential energy surface (PES) on which a path made of the local minima located on an arrangement continuity variable (named minimum energy path (MEP) coordinate and represented as a solid line in the plot of Fig. 1.2) connects reactants (A + BC) to products (AB + C);

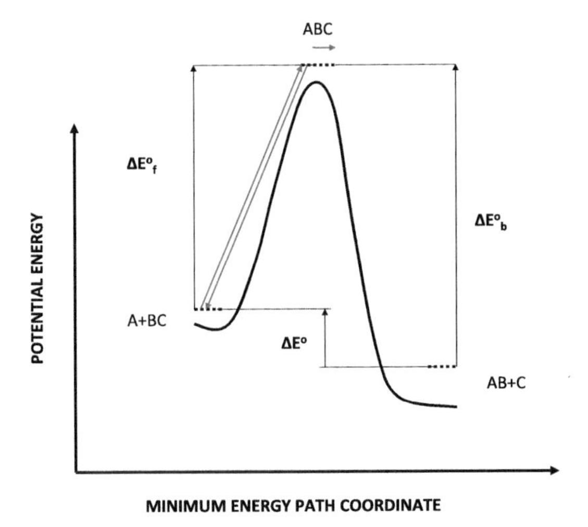

Fig. 1.2 A sketch of the transition state theory model for the exothermic reaction A + BC giving AB + C. The solid black line is the minimum energy path connecting reactants and products, which shows in its central part the transition state ABC (at the top) and the exothermicity ΔE^o (at the bottom). On the left-hand side (LHS), the forward process activation energy ΔE^o_f is shown, while the backward process activation energy ΔE^o_b is shown on the right-hand side (RHS)

2. On the MEP, one can locate an intermediate region supporting an ABC configuration (the transition state (TS)) separating reactants from products[1];
3. The TS population is in equilibrium with that of reactants according to relationship $\frac{Q^{\ddagger}_{ABC}}{Q_{BC}Q_A}e^{-\Delta E^o_f/k_B T}$ where ΔE^o_f is, indeed, the difference in energy internal between the TS and the reactants (often called activation energy) and Q_S is the partition function of the S system;
4. At the TS, the system crosses over (without recrossing back) to the products with a frequency $k_B T/h$;
5. The basic TST forward rate $k^{TST}(T)$ (hereinafter, we shall drop the label f when not strictly necessary) is set equal to the product of quantities associated with steps 3 and 4 giving

$$k^{TST}(T) = \frac{k_B T}{h}\frac{Q^{\ddagger}_{ABC}}{Q_{BC}Q_A}e^{-\Delta E^o_f/k_B T}. \tag{1.23}$$

Corrections to the basic TST rate can be introduced by formulating recrossing, tunneling, and steric effects corrections and by providing, as well, a more appropriate

[1]The TS is usually associated with a saddle. However, this association is an arbitrary assumption because (as it will be discussed in some detail in Chap. 4) the regions of the PES dividing trajectories back reflected from those crossing over are associated with periodic orbits dividing the (potential energy) surface (PODS). PODS can be more than one, do not necessarily sit on a saddle, and can be recrossed by the system.

definition of its location and/or energetics (e.g., variational, including centrifugal barriers, etc.) on the MEP or even by including statistical considerations derived by model or reduced dimensionality dynamical calculations.

1.1.4 Toward Detailed Single-Collision Studies

It has to be stressed here that the TST approach has no predictive power. Accordingly, its most popular use is as a phenomenological (empirical) equation whose coefficients are treated as best-fit parameters. As a matter of fact, the TST formulation of the rate coefficient is extensively employed as a practical way of implementing the integration of kinetic equations of the chemical subsystem of multiscale (atmospheric, combustion, etc.) simulations, the management of knowledge contents, and in artificial intelligence procedures.

Fortunately, in the last half century, both experimental and computational technologies have progressed enormously. For this purpose, it is worth recalling here that, before applying any model or accurate treatment to reactive processes, it should be taken into account that, for a thermalized system occurring on a single PES, the overall $k(T)$ results from statistically weighted (depending on the temperature of interest) sum of the detailed i (initial) to f (final) state contributions $k_{if}(T)$ (whose internal energies are ϵ_i and ϵ_f, respectively) as follows:

$$k(T) = \sum_i \sum_f w_i \frac{e^{[-\epsilon_i/k_B T]}}{Q_{BC}(T)} k_{if}(T). \tag{1.24}$$

The state-to-state rate coefficients $k_{if}(T)$ can be formulated in terms of the state-to-state cross section $\sigma_{if}(E_{tr})$ as follows:

$$k_{if}(T) = \int_0^\infty \sigma_{if}(E_{tr}) g(E_{tr}) \mathrm{d}E_{tr}, \tag{1.25}$$

where E_{tr} is the translational energy of the system and $g(E_{tr})$ is the translational energy distribution. For a gas in thermal equilibrium at temperature T, the function $g(E_{tr})$ has the form

$$g(E_{tr}) = \left(\frac{1}{\pi\mu}\right)^{1/2} \left(\frac{2}{k_B T}\right)^{3/2} E_{tr} e^{-E_{tr}/k_B T}. \tag{1.26}$$

Accordingly, by substituting (1.26) into (1.25), one has

$$k_{if}(T) = \left(\frac{1}{\pi\mu}\right)^{1/2} \left(\frac{2}{k_B T}\right)^{3/2} \int_0^\infty E_{tr} e^{-E_{tr}/k_B T} \sigma_{if}(E_{tr}) \mathrm{d}E_{tr}. \tag{1.27}$$

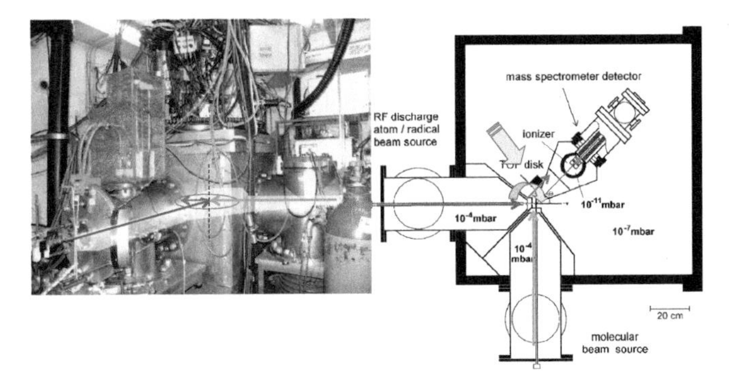

Fig. 1.3 LHS: a picture of the CMB apparatus built at the University of Perugia (IT); RHS: a draft of its technical scheme

Along this line, a big leap forward in the study of chemical reactions has been represented by the assemblage of crossed molecular beam (CMB) experimental apparatuses of which a sketch of the machine built at the University of Perugia (IT) is given in Fig. 1.3. As illustrated by the RHS of the figure, the apparatus consists of a high vacuum chamber into which the beams of the two reactants species are injected and collimated to intersect. Thanks to the high vacuum, the reactant molecules can collide only once and the products are measured in mass and direction by a rotating mass spectrometer. Their speed is instead evaluated from the time they take to reach the detection point.

The outcome of CMB apparatuses is a set of highly informative data on mainly bimolecular collisions providing a wealth of information on the

- primary reaction products,
- reaction mechanisms,
- structure and lifetime of transients,
- internal energy allocation of products, and
- detailed nature of the interaction,

that can be obtained (especially by coupling the CMB technology with laser ones which can not only state selectively operate on reactants and products but can also interact with transient species) through computations. Compute technologies are, in fact, the other vital ingredients of such studies because through heavy computations (that are becoming increasingly popular, thanks to the high-performance and high throughput features of modern platforms), it is becoming routinely feasible to work out a computational estimate (virtual experiment) of the measured properties of the real experiment and fully and accurately probe the interactions of the systems under consideration.

The interactions coming into play in molecular processes are, indeed, a specific feature of chemical reactions studies because of the richness of the variety of forces involved. As a matter of fact, such a variety is generated by the interplay of ionic and covalent, permanent and induced, short and long range, and two and many body interactions.[2] In order to illustrate most of the concepts developed by Molecular Science in its approach to "understanding" the nature of chemical processes, we shall focus in the following on the two particles (two bodies) before undertaking the study of more complex systems. The greater simplicity of this study will allow us to approach in a smoother way the basic concepts of chemical transformations and of the related computational approaches.

1.2 Classical Mechanics of Two-Particle Collisions

1.2.1 Reference Frame and Elementary Interactions

At present, we assume that the two bodies are represented as two points or spherical objects with masses in the three-dimensional physical space (after all a large part of the collision process takes place at distances at which the structure of the colliding bodies has hardly any effect) and that the behavior of such systems can be described by the laws of classical mechanics (which is a reasonable starting point for most of the dynamical computational chemistry applications).

In an arbitrary laboratory fixed reference system $(X, Y, Z)_{lab}$ (thick arrows of Fig. 1.4), according to classical mechanics, a system formed by two colliding particles A and B is uniquely defined in physical space by the two position vectors $\mathbf{r}_A$ and $\mathbf{r}_B$ and the two momentum vectors $\mathbf{p}_A$ and $\mathbf{p}_B$ defined, respectively, as $m_A\mathbf{v}_A$ and $m_B\mathbf{v}_B$ with m_i being the mass and $\mathbf{v}_i$ the velocity of either particle A or B).

These vectors are usually referred to as laboratory (*lab*) axis frame $(X, Y, Z)_{lab}$, a frame of Cartesian orthogonal X, Y, and Z axes having a fixed origin and orientation like those of the physical laboratory (in the following, we shall omit the specification *lab* when not strictly necessary). The position vectors $\mathbf{r}_A$ and $\mathbf{r}_B$ of the particles A and B, respectively, can be represented either in terms of their projections X_A, Y_A, Z_A, and X_B, Y_B, Z_B over the X, Y, and Z axes (not shown in Fig. 1.4 for the sake of simplicity) or in terms of the corresponding spherical polar coordinates (i.e., the moduli r_A and r_B of the two vectors $\mathbf{r}_A$ and $\mathbf{r}_B$ and the respective angles Θ_A, Φ_A and Θ_B, Φ_B.[3] Similar representations can be adopted for the vectors $\mathbf{p}_A$ and $\mathbf{p}_B$.

A representation of the two-body system isomorphous with the above-described one can be obtained using $\mathbf{r}_{CM}$, the position vector of the center-of-mass (CM) of the system, and $\mathbf{r}_{AB}$ (or $\mathbf{r}$ for short), the position vector of particle B with respect to particle A.

---

[2] Ionic is the interaction between charged particles (ions) in which the number of positive components (e.g., protons) differs from that of negatively charged particles (e.g., electrons). Covalent is the interaction associated with evenly shared particles (e.g., two atoms equally sharing the electrons). Permanent is a stable feature of the particles (e.g., the dipole moment). Induced is a temporary feature associated with the presence of another particle. Short and long ranges refer to the distance between the particles.

[3] The spherical polar coordinates of the particle i make use of r_i (the module of the position vector $\mathbf{r}_i$) and of its orientation angles Θ_i and Φ_i.

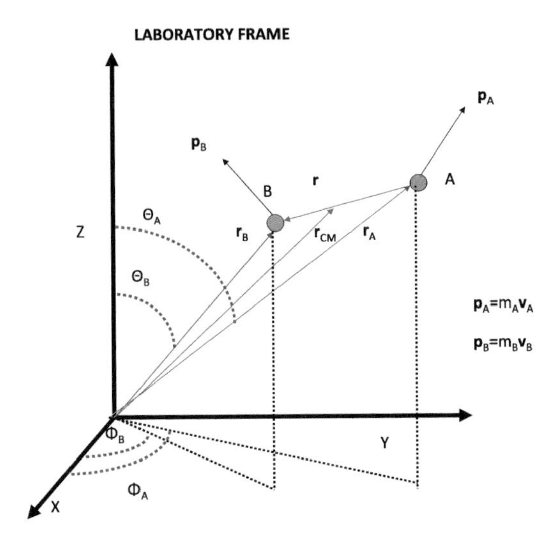

Fig. 1.4 LABORATORY FRAME: The vectors $\mathbf{r}_A$ and $\mathbf{r}_B$ define the position of the A and B colliding bodies with respect to the origin of the $(X,Y,Z)_{lab}$ frame. Related angles Θ_A, Φ_A, Θ_B, and Φ_B of the corresponding polar coordinates are also given. The related momentum vectors are $\mathbf{p}_A = m_A\mathbf{v}_A$ and $\mathbf{p}_B = m_B\mathbf{v}_B$ with $\mathbf{v}_i = \dot{\mathbf{r}}_i$. CM FRAME: The vectors $\mathbf{r}_{CM}$ and $\mathbf{r}$ define, respectively, the position of the CM with respect to the origin of the coordinate system and the position of particle B with respect to A

The position vector $\mathbf{r}$ is represented separately in Fig. 1.5 (LHS panel) using the $(x,y,z)_{CM}$ frame. It is worth pointing out here the use of small letters for the CM reference frame (as opposed to the capital ones used for the *lab* reference frame), of the spherical polar angles ϑ and ψ, and of the origin coinciding with the CM. The CM frame may have an arbitrary orientation (usually defined in terms of the values of angles α, β, γ (named Euler angles) by which the *lab* frame orientation needs to be (continuously) rotated so as to coincide with the plane defined by the position vector $\mathbf{r}$ and its velocity $\dot{\mathbf{r}}$ (this frame is called body fixed (BF)) and the angle formed by $\mathbf{r}$ and the BF z axis (not shown here) is called deflection angle θ).

In the same figure, we show in the upper RHS corner (using a $(X, Y, Z)_{lab}$ frame representation) a three-point stroboscopic picture (screenshots) of the $\mathbf{r}_A$ (dashed-dotted line) and $\mathbf{r}_B$ (dotted line) position vectors of the two bodies during a coplanar repulsive collision (related momenta are given as bold arrows and $\mathbf{r}$ values are given as dashed lines). In the lower RHS corner, the $\mathbf{r}$ values are represented separately in the sequence of occurrence.

A simplified illustration of the simplest cases of two-body interactions is given in Fig. 1.6: repulsive (central row) in which the trajectories of the two particles diverge and attractive (lower row) in which the trajectories of the two particles converge for a central potential depending only on the distance r ($r = |\mathbf{r}_A - \mathbf{r}_B|$) of the two bodies. Also shown in the figure is the case of no interaction (upper row), for a central potential depending only on the distance r ($r = \|\mathbf{r}_A - \mathbf{r}_B\|$) of the two bodies.

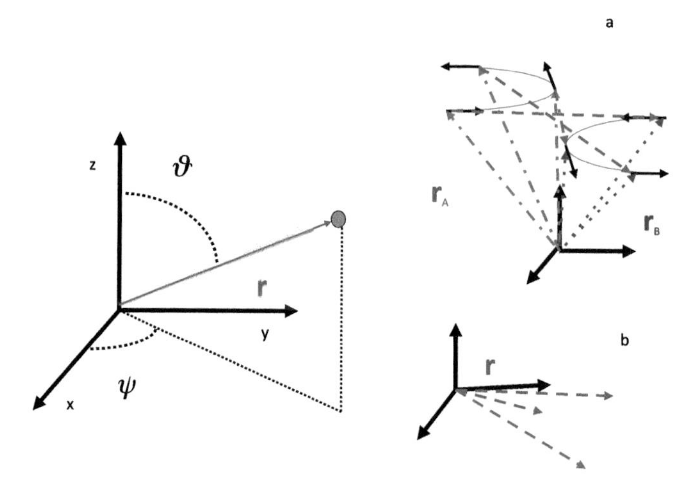

Fig. 1.5 Top RHS panel: a three-shot stroboscopic view of the coplanar A + B repulsive collision trajectory (two thin-solid lines occurring on a plane parallel to the X,Y axes of the (X, Y, Z)$_{lab}$ frame) complete of the relative three $\mathbf{r}_A$ (dashed dotted) and $\mathbf{r}_B$ (dotted) shots and of the associated three velocity vectors (bold arrows). The three shots of vector **r** are also shown as dashed lines. LHS panel: the vector **r** represented in its (x,y,z)$_{CM}$ reference frame. Lower RHS panel: the three shots of **r** shown in their sequence of occurrence in the (x,y,z)$_{CM}$ reference frame

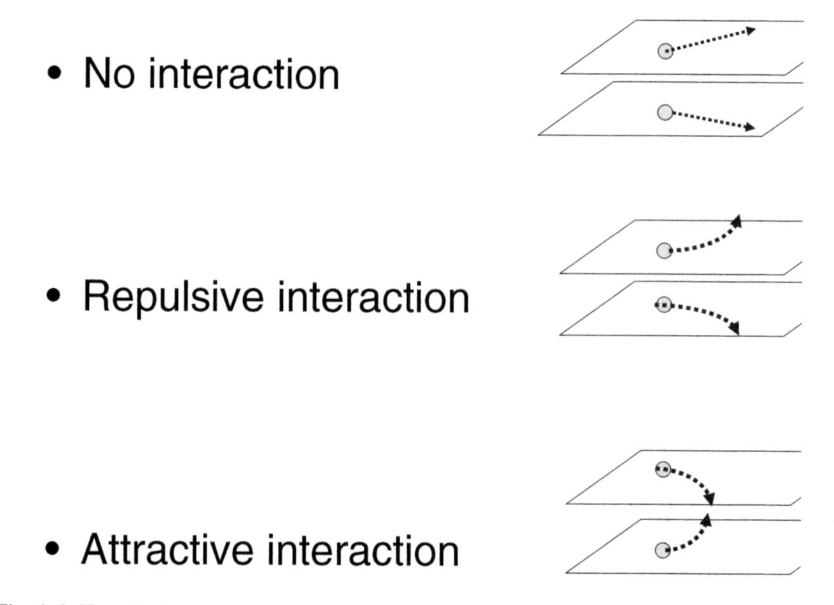

Fig. 1.6 Hypothetical trajectories of two bodies placed on the two parallel planes of their initial position and velocity. Top panel: no interaction, central panel: repulsive interaction, lower panel: attractive interaction

The repulsive interaction (like the one between two charges of the same sign) is completely intuitive. The repulsion is very large at small distance (at the distance where the particles nearly coalesce) and, as the two bodies move away from one another, the interaction monotonically decreases. Similarly, the second type of force attraction (such as that between two charges of opposite sign) is also negligible at large distances. The attractive force tends to bring the two bodies together and the magnitude of the interaction monotonically increases as they approach each other. The same reasoning applies to the multipoles with different orientations. The interaction is repulsive if the polarities are facing mainly of the same sign, and attract each other if the polarities are facing mainly of different signs.

Yet, the most important role in chemistry is played by the third type of interaction that can change from attractive to repulsive at different internuclear separations. It is less intuitive but even more general and realistic. This type of interaction tends to be more attractive at large distances and then becomes more repulsive at smaller internuclear distances. Thus, this system has an equilibrium distance where the interaction is most attractive.

1.2.2 The Equations of Motion

As we have just seen, a complete description of the system of two particles A and B, with masses m_A and m_B, must be based not only on the position vectors $\mathbf{r}_A$ and $\mathbf{r}_B$ but also on their variation over time. For this reason, it is necessary to determine the (linear) momenta $\mathbf{p}_A$ and $\mathbf{p}_B$ of the two particles as well. The space defined by the set of pairs of conjugated variables (position $\mathbf{r}_i$ and momentum $\mathbf{p}_i$ for all particles of the system under consideration is called "phase space"). Different points of the phase space are characterized by different states of a classical system (the position vectors $\mathbf{r}_A$ and $\mathbf{r}_B$ and their respective momenta $\mathbf{p}_{r_A}$ and $\mathbf{p}_{r_B}$). In Newton's second law, the motion of each particle is described by coupled 6N (where N is the number of particles of the system) mathematical equations $\mathbf{F}_i = \mathrm{d}\mathbf{p}_i/\mathrm{d}t$ and $\mathbf{p}_i = \mathrm{m}_i\mathbf{v}_i = \mathrm{m}_i\mathrm{d}\mathbf{r}_i/\mathrm{d}t$ in the coordinates of the chosen frame. The Hamiltonian of the two-body system is given by

$$\mathcal{H} = \mathcal{T} + \mathcal{V} = \sum_{i=A,B} \frac{\mathbf{p}_{r_i}^2}{2m_i} + V(\mathbf{r}_A, \mathbf{r}_B) = E, \quad (1.28)$$

where E is the total energy, $\mathcal{T}$ is the kinetic energy, and $\mathcal{V}$ is the potential energy (that in our case is $V(\mathbf{r}_A, \mathbf{r}_B)$). The energy E is a constant when we are considering a conservative system. The Hamiltonian $H = E$ for this system along with the initial conditions determines the fate of the system for all times (completely deterministic). As we shall see later, the positions and momenta can be determined by solving Hamilton's equations of motion, for each atom i.

In the laboratory frame, the system may be described using either an X, Y, Z Cartesian coordinate representation or, alternatively, other coordinates. The time dependence of the vectors $\mathbf{p}_{r_i}$ and $\mathbf{r}_i$ can be obtained using different classical formulations. Instead of the just mentioned popular Newton's formulation, we shall use the formulation of Hamilton[4] where one has a pair of first-order ordinary differential equations of the conically conjugated variables $p_{r_{iW}}$ and r_{iW} (with W = X,Y,Z being the set of chosen orthogonal coordinate system)

$$\frac{\mathrm{d}p_{r_{iW}}}{\mathrm{d}t} = -\frac{\partial \mathcal{H}}{\partial r_{iW}} \quad \text{and} \quad \frac{\mathrm{d}r_{iW}}{\mathrm{d}t} = \frac{\partial \mathcal{H}}{\partial p_{r_{iW}}} \tag{1.29}$$

for a total of twelve equations. These equations can be integrated numerically with standard techniques which we will mention later. In only a few special cases will the equations of motion (1.29) have analytical solutions. In the vast majority of cases, no analytical solutions are known. As we shall see later, analytical solutions, when they are available, have been generated only after performing laborious analytical transformations (yet, they have the advantage of allowing useful decompositions of the problem leading to both interesting insights and significant reduction of dimensionality of the problem). Very often, however, accurate approximations to the solution can only be found using numerical techniques.

In the case of conservative systems, the potential V depends only on the relative distance between the two particles. Then, the Hamiltonian of the system assumes a form particularly convenient when one uses CM coordinates $\mathbf{r}_{CM}$ and the internal coordinates $\mathbf{r}_{AB}$ (or more frequently simply $\mathbf{r}$). Their definition is immediate:

$$\mathbf{r}_{CM} \equiv \frac{m_A}{M}\mathbf{r}_A + \frac{m_B}{M}\mathbf{r}_B \quad \text{with} \quad M \equiv m_A + m_B \tag{1.30}$$

$$\mathbf{r}_{AB} \equiv \mathbf{r} = \mathbf{r}_A - \mathbf{r}_B. \tag{1.31}$$

Expressing the Hamiltonian (1.28) in these coordinates, we have

$$\mathcal{H} = \frac{\mathbf{p}^2_{r_{CM}}}{2M} + \frac{\mathbf{p}^2_{r_{AB}}}{2\mu} + V(r) \quad \text{with} \quad \mu = \frac{m_A m_B}{m_A + m_B}. \tag{1.32}$$

The first term of this Hamiltonian describes the motion of the CM. The second and the third terms in the Hamiltonian describe instead the relative motion of the two particles. Since $V(r)$ is independent of $\mathbf{r}_{CM}$, the motion of the CM relative to the laboratory system $(X, Y, Z)_{lab}$ is that of a free particle which moves at a constant velocity (inertial system). The CM coordinate system $(x, y, z)_{CM}$ differs from the laboratory one in that it is not fixed in space but moves at a constant velocity with respect to the laboratory coordinate system $(X, Y, Z)_{lab}$. Accordingly, the CM coordinate system $(x, y, z)_{CM}$ shown in Fig. 1.5 can also be used as the origin of an

---

[4]Another popular formulation of the equations of motion is the Lagrange's one

$$\frac{\mathrm{d}}{\mathrm{d}t}\frac{\partial L}{\partial \dot{r}_W} - \frac{\partial L}{\partial r_W} = 0,$$

where $L = \mathcal{T} - \mathcal{V}$ is the Lagrangian of the system.

inertial coordinate system. The CM set of coordinates is also said to be barycentric (center-of-gravity). The inertial system has the important property of bearing a null total linear momentum.

As described in the related caption, the RHS panels ((a) and (b) of Fig. 1.5) compare a collision of two equal masses occurring on the X,Y plane (as we shall see later this is a collision under the effect of a repulsive interaction) by giving the evolution of the $\mathbf{r}_A$ and $\mathbf{r}_B$ pair in a Lab frame (panel a) and the evolution of $\mathbf{r}$ in a CM frame having the same orientation as the Lab one in (panel b). The latter illustrates graphically the reduction of complexity associated with the decomposition of the $\mathbf{r}_A$ and $\mathbf{r}_B$ problem into a $\mathbf{r}$ and a $\mathbf{r}_{CM}$ problem, thanks to the introduction of conservation laws (in our case, the conservation of CM momentum). The two-center problem is, in fact, transformed into a one-center problem of a particle of mass μ, equal to the reduced mass of the system, subject to the potential or interparticle interaction as shown by the use of the only vector $\mathbf{r}$. Accordingly, the number of Hamilton's equations to be integrated is reduced from twelve to six, those relating to the three Cartesian components of the position vector $\mathbf{r}$ (r_x, r_y, r_z) (or their respective components of the polar representation r, ϑ, ψ (see Fig. 1.5)) and the three components of its conjugated momentum $\mathbf{p}_{r_{AB}}$.

1.2.3 *The Deflection Angle θ*

For the central field problem under consideration, it is possible to further decompose the problem using symmetry properties of the system. As already mentioned, in fact, in the case of the central field, the interaction potential depends only on the magnitude of the coordinate $\mathbf{r}$ and is preferable for convenience and clarity to use systematically, as we already do, $\mathbf{r}$ and $\mathbf{p_r}$ instead of $\mathbf{r}_{AB}$ and $\mathbf{p}_{\mathbf{r}_{AB}}$ for the diatomic variables. For the same reasons, we also assume that $V(r) \to 0$ when $r \to \infty$ (except when explicitly said), by setting the zero of energy to the asymptotic value of the potential. Then by choosing, as is done in Fig. 1.7, the orientation of the axes of the system of reference so that two of them (for example, z and y) lie in the plane determined by the initial velocity vector, the system will remain confined to the simple trajectory ($\mathrm{d}\psi/\mathrm{d}t = 0$), since the potential depends only on the magnitude of the vector $\mathbf{r}$. Following this transformation, the classical Hamiltonian can be written explicitly in the following way[5]:

[5] In fact, see Fig. 1.7, we have for the components (z, y) of $\mathbf{r}$, $z = -r\cos\theta$ and $y = r\sin\theta$ ($\theta - \pi/2 = \vartheta$) or by differentiating with respect to time

$$v_z \equiv \frac{\mathrm{d}z}{\mathrm{d}t} = -\dot{r}\cos\theta + r\frac{\mathrm{d}\theta}{\mathrm{d}t}\sin\theta \;\; \textit{and} \;\; v_y \equiv \frac{\mathrm{d}y}{\mathrm{d}t} = \dot{r}\sin\theta + r\frac{\mathrm{d}\theta}{\mathrm{d}t}\cos\theta.$$

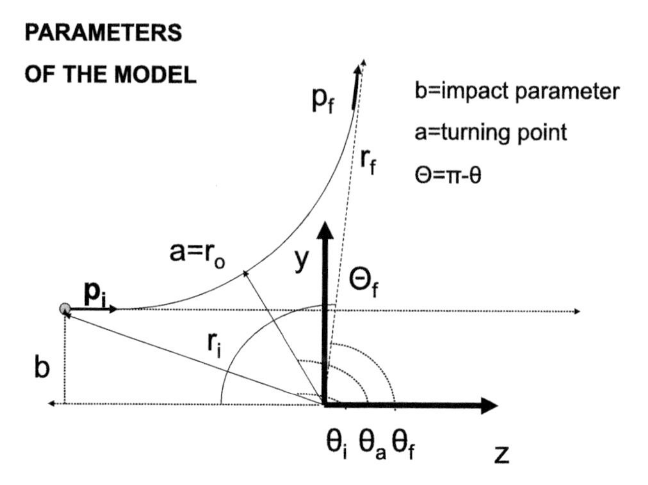

Fig. 1.7 Hypothetical trajectory of a system of reduced mass μ, impact parameter b, and momentum $\mathbf{p} = \mu \mathbf{v_r}$ for a repulsive interaction. The angle Θ is the scattering angle, while the angle θ (subindices i, a, and f mean initial, at the turning point and final, respectively) is the deflection angle, r_o is the value of classic reversal of r (minimum distance a=r_o or point of inflection also called turning point)

$$E = \mathcal{H} = \mathcal{T} + \mathcal{V} \qquad \begin{cases} \mathcal{T} = T = \frac{1}{2}\mu v^2 = \frac{1}{2}\mu(v_z^2 + v_y^2) = \frac{1}{2}\mu(\dot{r}^2 + r^2\dot{\theta}^2) \\ \mathcal{V} = V(r) \end{cases} \tag{1.33}$$

(with v we denote the relative velocity $\dot{r} = \mathrm{d}r/\mathrm{d}t$ having the initial v_0 at time $t = 0$) and the Hamilton equations to be integrated are further reduced from six to four.

The system is conservative and the Lagrangian in polar coordinates for this central field problem is

$$L = T - V = \frac{\mu}{2}(\dot{r}^2 + r^2\dot{\theta}^2) - V, \tag{1.34}$$

and therefore the conjugated momentum to θ is

$$p_\theta = \frac{\partial L}{\partial \dot{\theta}} = \mu r^2 \dot{\theta} \tag{1.35}$$

and the conjugated momentum to r is

$$p_r = \frac{\partial L}{\partial \dot{r}} = \mu \dot{r}. \tag{1.36}$$

The Hamiltonian expressed in terms of the conjugated variables reads

$$H = T + V = \frac{p_r^2}{2\mu} + \frac{p_\theta^2}{2\mu r^2} + V = E \tag{1.37}$$

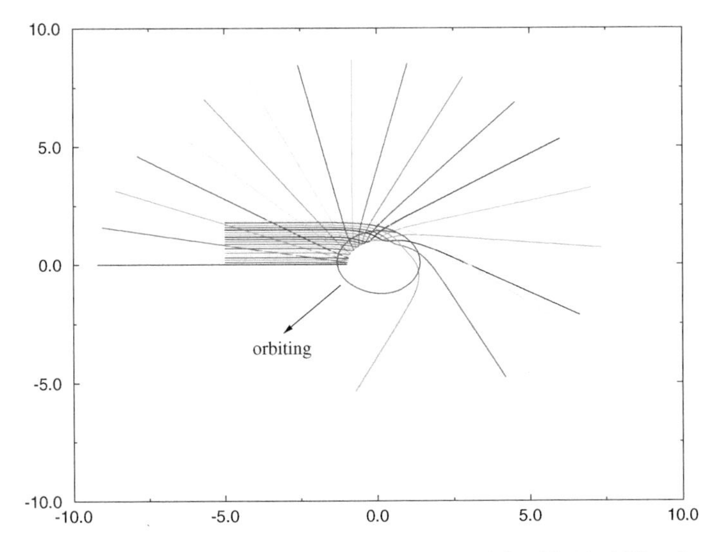

Fig. 1.8 Trajectories for the Lennard–Jones (6–12) potential (defined in Eq. 1.73 at the end of the present chapter) having an impact parameter b varying between 0 and 2 in steps of 0.08 reduced units $b = b^*/\sigma$ (the formula of the potential and the significance of the parameter σ will be discussed later). In the figure, the particular case that leads to an orbiting trajectory is also shown

with E being as usual the total energy. Accordingly, the Hamilton equations of motion become

$$\dot{\theta} = \frac{d\theta}{dt} = \frac{\partial H}{\partial p_\theta} \quad \text{and} \quad \dot{p}_\theta = \frac{dp_\theta}{dt} = -\frac{\partial H}{\partial \theta} \tag{1.38}$$

$$\dot{r} = \frac{dr}{dt} = \frac{\partial H}{\partial p_r} \quad \text{and} \quad \dot{p}_r = \frac{dp_r}{dt} = -\frac{\partial H}{\partial r}. \tag{1.39}$$

The set of cartesian $z(t)$ and $y(t)$ or polar $r(t)$ and $\theta(t)$ coordinates that are obtained by integrating the Eq. 1.29 for the former ($W = z, y$) or (1.38) and (1.39) for the latter ($W = r, \theta$) defines the trajectory followed by the particle (see again Fig. 1.7). Initial values of the cartesian coordinates z_i and y_i are easy to set: z_i to a sufficiently large negative value and $y_i = b$; initial values of related linear momenta p_{z_i} and p_{y_i} are also easy to set: $p_{z_i} = \sqrt{2\mu E}$ and $p_{y_i} = 0$. Slightly more involved is the determination of the initial conditions when using polar coordinates: r_i is set to a sufficiently large value and $\theta_i = \pi - \arcsin b/r_i$; initial values of related linear momenta p_{r_i} and p_{θ_i} are $p_{r_i} = \sqrt{2\mu E}\cos\theta_i$ and $p_{\theta_i} = r\sqrt{2\mu E}\sin\theta_i$ where we have assumed that the asymptotic value of the potential energy is zero.

By way of example in Fig. 1.8, we show several trajectories (each has a different impact parameter but the same initial relative speed) followed by a particle moving under the effect of the attractive–repulsive (attractive at large distances and repulsive at small distances) Lennard–Jones (6,12) model potential (commonly used to formulate dispersion van der Waals interaction) that will be discussed in more detail later on. The trajectories were obtained by integrating Eqs. 1.38 and 1.39 using a numerical method.

Before addressing the question of the methods used for numerical integration, we discuss here the further simplification of the equations for the simple (though very general and important) problem of the deflection from the central field. This problem can in fact be further decomposed (thereby reducing from 4 to 2 the number of differential equations to integrate), thanks to the use of conservation laws related to total angular momentum **L** and the total energy E. For this purpose, use is made of the magnitude of the impact parameter b defined as the distance between the center of interaction and the initial velocity vector (or, equivalently, the perpendicular segment drawn from the particle to the z axis, see Fig. 1.7). In many treatments of diatomic molecules, one uses **j** as the angular momentum operator. For the interaction of two spherically symmetric particles, $\mathbf{L} \equiv \mathbf{j}$. To be consistent with the scattering notation used later on, we will use **L** as the total angular momentum when we consider diatomic molecules. In fact we will always use **L** for systems where the position vector **r** associated with with $\mathbf{L} \equiv \mathbf{r} \times \mathbf{p_r}$ is used to describe particles when are infinitely separated. In the classical formulation of the conservation of total angular momentum l as[6]

$$|\mathbf{L}| = \mu v_0 b = l = \mu r^2 \dot{\theta} \tag{1.40}$$

contrary to the quantum one[7] and combining Eqs. (1.33) and (1.40), one obtains the two differential equations that characterize the system. Of these, the radial velocity is

$$\dot{r} \equiv \frac{dr}{dt} = \pm \left[\frac{2}{\mu} \left(E - V(r) - \frac{l^2}{2\mu r^2} \right) \right]^{1/2} \tag{1.41}$$

(in the following, we shall often compact the notation using the effective potential $V^l(r)$ which incorporates the centrifugal component $l^2/2\mu r^2$ being $V^l(r) = V(r) + l^2/2\mu r^2$) and then the angular velocity is

$$\dot{\theta} \equiv \frac{d\theta}{dt} = \frac{l}{\mu r^2} = \frac{v_0 b}{r^2}. \tag{1.42}$$

---

[6]By definition of angular momentum and vector product is in fact:

$$\mathbf{L} = \mathbf{r} \times \mathbf{p} = \mu(\mathbf{r} \times \dot{\mathbf{r}}) = \mu \left(z\frac{dy}{dt} - y\frac{dz}{dt} \right).$$

in this case, the motion is confined to a plane yz as in our case. Then, at time $t = 0$, since $dy/dt = 0$, we have $|\mathbf{L}| = y\mu dz/dt = \mu b v$, while at generic times t, you will have $|\mathbf{L}| = \mu r^2 \dot{\theta}$ (see next note).

[7]In the quantum treatment in formulating the conservation of the diatomic total angular momentum quantum number l, l^2 will be replaced (apart from a constant factor) by $l(l+1)$ which is the total angular momentum eigenvalue of the quantum operator, as we will see more forward.

A further simplification of the calculation is obtained from the elimination of the time variable by dividing Eq. (1.42) by (1.41). We then obtain the relation between the variation of the angle and the variation of r

$$\frac{\mathrm{d}\theta}{\mathrm{d}r} = \frac{\dot{\theta}}{\dot{r}} = \pm br^{-2}\left[1 - \frac{b^2}{r^2} - \frac{V(r)}{E}\right]^{-1/2}. \tag{1.43}$$

By integrating Eq. (1.43) with respect to r from ∞ to the classical turning point a and then from a to ∞ for the second part of the collision (by symmetry this is equivalent to doubling its value computed by integrating from a to ∞), we get the following formulation of the deflection angle θ as a function of the parameters E, b, and potential $V(r)$

$$\theta = \pi - 2b\int_a^{\infty} r^{-2}\left[1 - \frac{b^2}{r^2} - \frac{V(r)}{E}\right]^{-1/2} \mathrm{d}r. \tag{1.44}$$

The angle θ_a is the angle of closest approach corresponding to the distance $r = r_0 = a$ (the distance of closest approach also called either classical turning point or point of return). In r_0 the radial velocity $\dot{r}$ is zero and the total energy E is equal to the effective potential energy (see Eq. 1.41). The value of a corresponds to the larger root of the quadratic equation given by $1 - b^2/r^2 - V(r)/E = 0$ or

$$r_0 = b\left[1 - \frac{V(r_0)}{E}\right]^{-1/2}. \tag{1.45}$$

It should be emphasized that knowledge of the dependence of the angle of deflection (or deflection function) on the parameters of the system allows one to derive all of the important properties of dilute gases. Examples are transport properties (see Ref. [1]) such as the viscosity and the second virial coefficient. Indeed, for the viscosity $\eta(T)$ of a dilute gas, one has

$$\frac{RT}{\eta(T)} = \frac{4}{5}N\sqrt{\pi}\int_0^{\infty} e^{-x^2}x^7\left[\int_0^{\infty} \sin^2\theta\, \mathrm{d}b^2\right]\mathrm{d}x, \tag{1.46}$$

and for the second virial coefficient $\mathrm{B}(T)$[8]

$$B(T) = \frac{4}{5}N\sqrt{\pi}\int_0^{\infty} e^{-x^2}x^4\left[\int_0^{\infty} \theta\, \mathrm{d}b^3\right]\mathrm{d}x, \tag{1.47}$$

[8]The equation state (or virial) is

$$pV_m/RT = 1 + B(T)/V_m + C(T)/V_m^2 + D(T)/V_m^3 \quad V_m = \text{molar volume.}$$

where $x^2 = E/k_BT$ and k_B is the Boltzmann constant ($k_B \approx 1.38066 \times 10^{-23}$ J/K). In a dilute gas, in fact, the dominant component is the two-body interaction. Three- and more-body interactions are not important until the gas is very dense or condenses to form a liquid or solid where many-body interaction become very important.

1.3 The Computation of Scattering Properties

1.3.1 Trajectories integration (Hamilton equations)

If you want to know the detailed temporal evolution of the system, the equations of motion (1.38) and (1.39) must be integrated numerically. A very simple numerical method for integrating first-order differential equations is based on the Euler method which uses an approximation of the first derivative by the quotient of the function $f(x)$ at neighboring points and the distance between those points

$$\frac{\mathrm{d}}{\mathrm{d}x} f(x) \simeq \frac{\delta f(x)}{\delta x} = \frac{f(x_{i+1}) - f(x_i)}{h} \tag{1.48}$$

for the generic point i with stepsize $h = \delta x = x_{i+1} - x_i$. From Eq. (1.48), we obtain the formula

$$f(x_{i+1}) = f(x_i) + h \left[\frac{d}{dx} f(x) \right]_{x=x_i} + O(h^2). \tag{1.49}$$

Thus, by knowing $f(x_i)$ and its first derivative $\frac{d}{dx} f(x)|_{x=x_i}$, one can calculate the value of f at the next point x_{i+1}.

The price paid in exchange for the simplicity of this formula is to accumulate at each step an error of first order (which is conventionally referred to as $O(h^2)$, where h is the integration step) that is too large to allow further integration for sufficiently long intervals. The related algorithm has the following structure when using the polar coordinate formalism:

```
PROGRAM TRAJECTORY BY INTEGRATION OF HAMILTON EQUATIONS
INPUT am1, am2, ro, b, etr, tstep, tsup
FUNCTIONS pot, dpot
---------------------------------------
am1    is the mass of particle 1
am2    is the mass of particle 2
ro     is the initial distance between the two particles
b      is the impact parameter in angstroms
etr    is the initial translational energy in kcal/mol
potr   is the potential function with radial distances
           in angstroms and energies in kcal/mol
dpotr  is the derivative of the potential with respect to r
th     is the value of theta
po     is the initial value of the momentum
```

```
tstep is the value of the time step used for the integration
tsup  is the maximum time
-----------------------------------------------------
r = ro
th = acos(-1.) -asin(b/r)
rmass = am1*am2/(am1 + am2)
po = sqrt(2.*rmass*etr)
pr = po*cos(th)
th_dot=po*b/r/r/rmass
time REPEAT FOR GOING FROM 0 TO PAST tsup by tstep
     r = r + pr/rmass*tstep
     th = th+th_dot*tstep
     pr =  b^2/(rmass* r^3) - dpotr(r)
     PRINT ' value of r and theta ', r, th
     IF (r> ro) END
END REPEAT time
```

It should be noted here that the approximation used for the first derivative is the forward difference, made using the value of the variable x and the next point. Similarly, one can evaluate the approximate derivative by a backward difference, i.e., between the point being considered and the previous point. By combining these two methods, one can develop more accurate and efficient algorithms. For higher accuracy, one can use multistep algorithms (see Ref. [2]).

1.3.2 Numerical Computation of θ

In general, except for the few cases (some of which, because of their widespread use in many applications of molecular modeling, will then be considered explicitly) in which the potential has a formulation that allows you to express the closed-form solutions, also the estimate of deflection angle θ has to be carried out using numerical quadrature.

To evaluate the integral that appears in the formula (1.44), some precautions have to be used. In fact, at the lower extreme (that can be easily determined by finding the lower root of the denominator), the integrand is singular. Furthermore, the integral is open to the right.

Before taking care of these two important features, let us consider its closed approximation obtained by cutting the integral at a given value b of the integration variable r. One of the simple methods which terminates the integral at large distance is the trapezoidal rule. The trapezoidal formula approximates the value of the finite integral $\int_a^b f(x)dx$, closed both left and right, as the sum of the area of the n trapezoids formed by the same number of values sampling the integrand $f(x)$ and the range of x to which it refers. The trapezoidal approximant has the form

$$\int_a^b f(x)dx \approx \sum_{i=1}^{n-1}(x_{i+1} - x_i)\frac{f(x_{i+1}) + f(x_i)}{2}) \approx hx\sum_{i=1}^{n-1} f(x_i + \frac{h}{2}), \quad (1.50)$$

in which $x_1 = a$ and n is sufficiently large to make $V(x)$ negligible. One should note that this final expression evaluates f at $x_i + \frac{h}{2}$ and thus misses the classical turning point or singularity. The approximation used in the rightmost formulation in (1.50) refers to the case where we consider the trapezoid formed by the value of the function $f(x)$ at the midpoint of the interval and constant step intervals (of size $h = x_2 - x_1 = ... = x_{i+1} - x_i$). The trapezoidal method, per se, is also inadequate because it considers an arbitrarily fixed upper limit (x_{sup}). The pseudocode is, however, as follows:

```
PROGRAM CMT1D: CLOSED MIDPOINT TRAPEZOIDAL 1D
INPUT  x_inf,x_sup, n
-----------------------------------
x_inf is the lower limit of the integral
x_sup is the upper limit of the integral
n is the number of sampling points of the
   function in the interval x_inf to x_sup
f is the integrand function to be defined a
  f(r)=\sqrt(1-b*b/r/r-V(r)/E)/r/r in which E and b are given
  within f that is defined as a function of those parameters
  and of the potential energy function V(r)
-----------------------------------
sum = 0
dx = (x_sup - x_inf) / n
x_point =x_inf + 0.5 * dx
i_point REPEAT FOR GOING FROM 1 TO n-1
    sum = sum + f (x_point)
    x_point = x_point +  dx
END REPEAT i_point
val_integr = sum*dx
OUTPUT val_integr
```

Note in the pseudocode the choice of fixing the number of iterations using a repetitive structure up to n steps rather than using a conditional iteration REPEAT UNTIL structure in order to maintain control of the number of iterations is allowed. This has the advantage of facilitating the distributing, if wished, of the calculations on many processors. It should also be said that in the above pseudocode, all of the data is known (and therefore input), i.e., values of the lower bound x_{inf}, the upper bound x_{sup}, and the number of quadrature grid points n. Also in the integrand, it is assumed that other parameters such as energy E, the impact parameter b, and the value of the potential $V(r)$ are given by the $f(r)$. Obviously, in a more articulated code, some of these values may be determined differently (e.g., a numerical search of the most suitable value of a or the search for a particular value of b). This is the case, in particular, when dealing with the study of the physical problem of the collision of two bodies in which you can have all or part of these parameters varied by the program. For example, the closed integral could be evaluated using a simple Monte Carlo method. In that case, a finite area fully including a portion covered by the integral needs to be defined (see Ref. [2]). The Monte Carlo method, in fact, associates the estimate of the integral with the fraction of the finite area covered by the values of the integrand for randomly sampled values of the variables. This point is important since an estimate of the convergence can be obtained. In our case, the convergence is to be found both with respect to the number of points sampling the function within the interval of x of a single closed integration and sampling the value of the overall integral when adding more intervals.

The first convergence can be obtained by iterating over the density of points within the same interval (whose pseudocode is rendered below as the CCMT1D subroutine) still by adopting the midpoint trapezoidal method.

```
SUBROUTNE CCMT1D (x_inf, x_sup, n_max, tol): CONVERGED CLOSED
MIDPOINT TRAPEZOIDAL 1D
OUTPUT val_integr
------------------------------------
x_inf is the lower limit of the integral
x_sup is the upper limit of the integral
 is the number of sampling points of the
   function in the interval a to xsup
n_max is the maximum number of duplication times (must be
   larger than 1) of the sampling points of the function in the
   interval x_inf to x_sup
tol is the accepted tolerance in the difference between the
   value of the integral at two subsequent iterations
f is the integrand function to be defined a
  f(r)=\sqrt(1-b*b/r/r-V(r)/E)/r/r in which V(r) is defined
  as a function as a function of those parameters
  and of the potential energy function V(r)
------------------------------------
dx = x_sup - x_inf
n=1
x_point =x_inf + 0.5 * dx
val_prev=dx * f (x_point)
i_n REPEAT FOR GOING FROM 2 TO nmax-1
   sum=0
   n=2*n
   dx=dx/2
   x_point =x_inf + 0.5 * dx
   i_point REPEAT FOR GOING FROM 1 TO n-1
      sum = sum +  f (x_point)
      x_point = x_point +  dx
  END REPEAT i_point
  val_integr = sum*dx
  IF(abs(val_integr-val_prev)<tol)  RETURN val_integr
  val_prev=val_integr
END REPEAT i_n
TELL ' lack of convergence in the range '
```

Sampling points as a function of the initial range are doubled at each iteration and the integral evaluated to determine the difference between the estimate obtained in a given iteration and that obtained in the previous one are used to check convergence toward an accepted error tolerance. It is important to note here that the choice of doubling the points has the great advantage of increasing significantly the number of points without needing to recalculate all of them as it would be when increasing, and the number of points by 1 at each step.

The second type of convergence, external, is then needed in order to gradually extend the range by adding to the upper bound additional intervals. The associated iteration is closed when the values calculated in two successive iterations differ with minimal error. You will have, therefore, the following:

```
PROGRAM CORMT1D: CONVERGED CLOSED OPEN RIGHT TRAPEZOIDAL 1D
INPUT x_left, x_right, nprove, nintv, error
OUTPUT
---------------------------------------
x_inf is the lower limit of the integral
x_sup is the upper limit of the integral
nprove is the maximum number of doubling
      of the sampling points of the function
      in the range considered
nintv is the maximum number of intervals considered
error is the maximum tolerated error difference
      between values of successive iterations
f is the integrand function to be defined a
  f(r)=\sqrt(1-b*b/r/r-V(r)/E)/r/r in which V(r) is defined
  as a function as a function of those parameters
  and of the potential energy function V(r)
---------------------------------------
valintgrtot = 0
dx = ( x_left - x_right)/(nintv-1)
i_intv REPEAT FOR GOING FROM 1 TO nintv
  x_inf=x_left
  x_sup=x_right
  CALL SUBROUTINE CCMT1D (x_inf, x_sup,n_prove, error)
  IF(abs(valintgrt-val_integr).< error) RETURN valintgrtot
  valintgrtot=valintgrtot+ val_integr
  x_left=x_right
  x_right=x_left+dx
END REPEAT i_intv
TELL ' lack of convergence in the range '
```

1.3.3 Other Collisional Properties

As will be further commented later, other quantities important for the characterization of the interaction of two bodies are the radial component of the classical action $\Delta(E, L)$ and the collisional delay time $\tau(E, L)$. The radial component of the classical action (see Eq. 1.33) is defined as

$$\begin{aligned}\Delta(E, L) &= 2\left\{\int_a^\infty p(r)\mathrm{d}r - \int_b^\infty p^0(r)\mathrm{d}r\right\} \\ &= 2mv\left\{\int_a^\infty \left[1 - \frac{b^2}{r^2} - \frac{V(r)}{E}\right]^{1/2} \mathrm{d}r - \int_b^\infty \left[1 - \frac{b^2}{r^2}\right]^{1/2} \mathrm{d}r\right\},\end{aligned} \tag{1.51}$$

where $p^0(r)$ and $p(r)$ are the radial momenta of the system, respectively, in the absence and in the presence of the potential. $\Delta(E, L)$ fulfills a vital role in the classical description of the trajectories (periodic orbits, swings, etc.), and by requiring the discretization of the classical action, you can find a classical analogue of the quantum numbers (see Ref. [3]). Also, the classical action and its conjugated variable (the angle θ) are a pair of variables (called action-angle variables) that can either be or are actually used to describe the processes of scattering involving periodic motions and allows one to include semiclassical effects of resonance and interference. The collisional delay time is defined as the difference between the collision time calculated in the presence and absence of potential

$$\tau(E, L) \equiv \int_{with\ Potential} \mathrm{d}t - \int_{no\ Potential} \mathrm{d}t = 2\left\{\int_a^\infty \frac{\mathrm{d}r}{\dot{r}} - \int_b^\infty \frac{\mathrm{d}r}{\dot{r}}\right\}$$
$$= \frac{2}{v}\left\{\int_a^\infty \frac{\mathrm{d}r}{\sqrt{1 - \frac{b^2}{r^2} - \frac{V(r)}{E}}} - \int_b^\infty \frac{\mathrm{d}r}{\sqrt{1 - \frac{b^2}{r^2}}}\right\}. \quad (1.52)$$

The delay time can be positive or negative depending on the type of potential used. In the case of a purely repulsive potential, in fact, the particle remains in the field of force a shorter time than the free particle (zero potential). In the case where the potential is even partially (long-range) attractive, however, it will remain in the interaction region a longer time than a free particle. From the sign and magnitude of the delay time, we find interesting information about the collision processes since they allow us to estimate the average lifetime of any intermediate complex of the collision.

1.3.4 The Cross Section

So far we have only considered individual collision events. The observable is always the result of a measurement on a macroscopic scale for a large amount of initial conditions, energies, and impact parameters even if in some cases these are selected to be homogeneous. Some chemical processes occur in a vacuum, with selection of the initial states and the velocity distributions have been narrowed as practically feasible in molecular beam experiments. In the already-mentioned CMB experiment, in fact, two beams with state selected internal energies and the monochromatic velocities are made to collide and then monitored, around the collision point, with a narrow angular resolution and determination of the energies of the scattered particles. It has been pointed out that an observable of great importance that can be detected is the differential cross section $\sigma(\theta, E_{tr})$. The differential cross section is a function of both scattering angle θ and the collision energy E_{tr} (because in the case of molecular partners one can have state selected internal energies of both the reactants and the products).

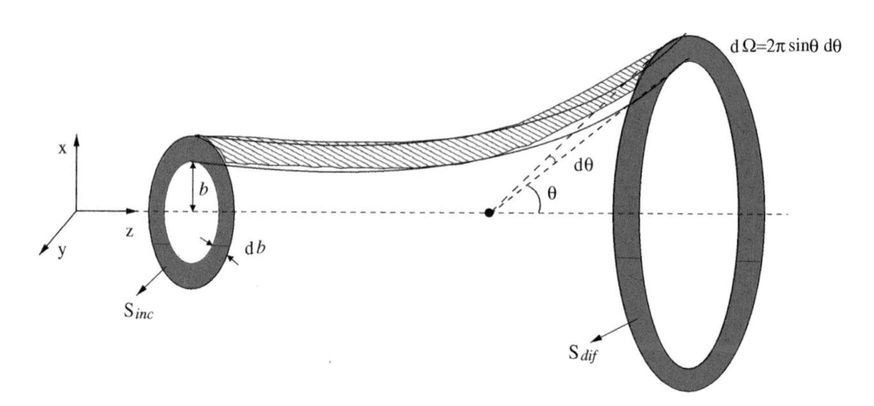

Fig. 1.9 Deflection of the differential element of the incoming flux ϕ_{in} (in the corona included in b and $b + db$ interval ($d\phi_{in} = 2N\pi b db$) dispersed in the area ($d\phi_{out} = 2N\sigma(\theta, E_{tr})\pi \sin\theta d\theta$))

It is obtained by equating the infinitesimal flux of incoming $d\phi_{in} = 2N\pi b db$ to the outgoing $d\phi_{out} = 2N\sigma(\theta, E_{tr})\pi sin\theta d\theta$ (see Fig. 1.9), where N is the total number of particles per second per unit area and $2\pi \sin\theta d\theta$ is an infinitesimal element of solid angle $d\Omega$. From the equality of the two flows is obtained the expression for the differential cross section[9]

$$\sigma(\theta, E_{tr}) = \left|\frac{2N\pi b db}{2N\pi \sin\theta d\theta}\right| = \left|\frac{b}{\sin\theta(d\theta/db)}\right|. \tag{1.53}$$

In cases where more than one value of b gives rise to the same scattering angle θ, the term on the right of the expression (1.53) is replaced, as follows, by a sum

$$\sigma(\theta, E_{tr}) = \sum_b \left|\frac{b}{\sin\theta(d\theta/db)}\right|. \tag{1.54}$$

The total elastic cross section, σ_{tot}, is derived from the differential cross section by performing an integration over θ

$$\sigma_{tot}(E_{tr}) = \int \frac{d\sigma}{d\Omega} d\Omega = 2\pi \int_0^{\pi} \sigma(\theta, E_{tr}) \sin\theta d\theta. \tag{1.55}$$

[9]From CMB experiments, it is impossible to distinguish between positive and negative deflections. Accordingly, the absolute value of θ is considered.

1.4 Popular Scattering Model Potentials

1.4.1 The Rigid Sphere Model

There is an extensive list of chemical models for the potential energy used to interpret and predict physical phenomena in the gas, liquid, and solid phases. The models that are commonly used in these cases attempt to represent the potentials in a simple yet realistic manner using a minimum number of parameters with a strong physical significance. In our case, the two particles may represent nuclei, atoms, electrons, fragments of the molecule, whole molecules, or still even more complex bodies. Therefore, even if there are theoretical means for improving the quality of these model potentials (see Chaps. 3 and 4), they are of widespread use in modeling chemistry.

A model potential that is simple to treat with a large number of applications in chemistry is the rigid sphere potential (which is repulsive). This model assumes that there is no interaction at all from d to infinity (in practice, the sum of the radius of the particles considered or, equivalently, the diameter of a sphere of interaction is impenetrable) to become infinitely repulsive at d (see Figure 1.10).

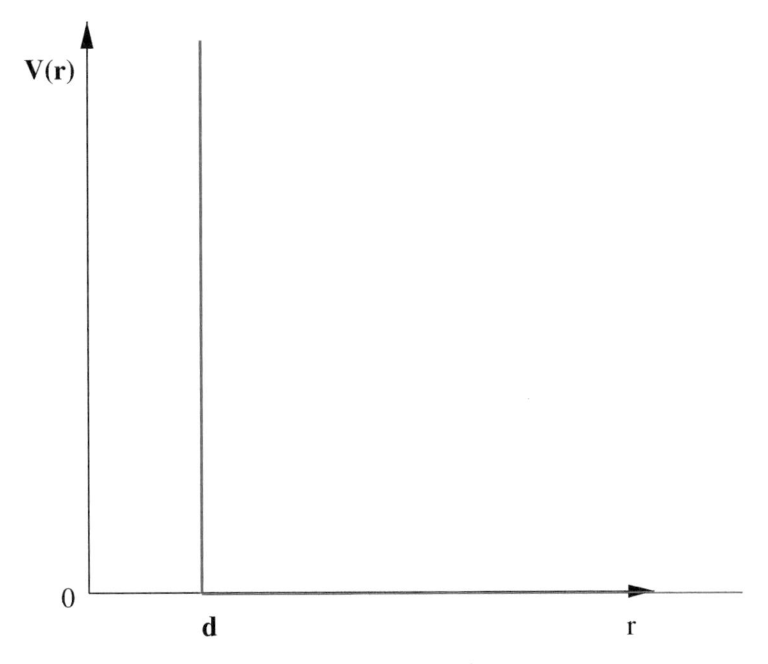

Fig. 1.10 The rigid sphere potential of radius d: $V(r) = \begin{cases} \infty & \text{if } r < d \\ 0 & \text{if } r \geq d \end{cases}$

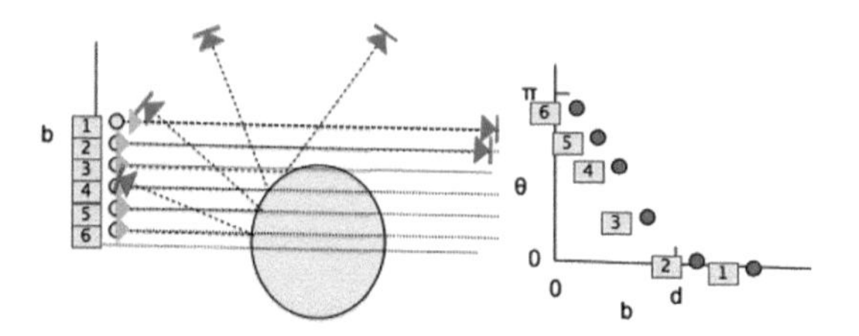

Fig. 1.11 LHS panel: Numbered trajectories (as from a power point file) of the rigid sphere potential of radius d computed by integrating the related Hamilton equations (see Eqs. 1.38 and 1.39); RHS panel: associated θ values plotted as a function of the impact parameter b

In this case, the classical turning point is

$$\begin{cases} a = d \quad if \quad b \leq d \\ a = b \quad if \quad b > d. \end{cases}$$

This is further illustrated in Fig. 1.11 where in the LHS panel the rigid sphere is represented by the circle. The figure shows also the various trajectories numbered from 1 to 6 in going from larger to smaller values of b. The corresponding computed values of θ (numbered accordingly) are plotted in the RHS panel of the figure. They coincide with related analytical solutions obtained as follows: Let $z = b/r$, Eq. (1.44) becomes

$$\begin{aligned} \theta &= \pi - 2b \int_0^{1/d} \left[1 - b^2 z^2\right]^{-1/2} dz \\ &= \pi - 2 \arcsin(b/d) = 2 \arccos(b/d) \quad to \ b \leq d \\ &= 0 \qquad to \ b > d, \end{aligned}$$

thanks to the use of the auxiliary variable $t = \cos z$. Accordingly, for the trajectories with $b \leq d$, we have

$$b = d \sin(\pi/2 - \theta/2) = d \cos(\theta/2) \tag{1.56}$$

and from Eq. (1.53), we obtain the differential cross section

$$\sigma(\theta, E_{tr}) = \frac{1}{4} d^2 \tag{1.57}$$

which has no preferential directions (the process is uniform in all directions and is independent of energy). Consequently, the total cross section will be

$$\sigma_{tot}(E_{tr}) = 2\pi \int_0^{\pi} \sigma(\theta, E_{tr}) \sin \theta d\theta = \pi d^2. \tag{1.58}$$

The value of the cross section πd^2 is the area of a circle or equivalently the maximum circumference of sphere with radius d.

It is also worth noting that this extremely simplified model of the interaction still allows one to derive the constant velocity Arrhenius rate coefficient $k(T)$. This model of the chemical reaction based on the rigid sphere potential assumes that collisions leading to contact of the two particles are reactive if the translational energy is greater than a threshold value E_a. So nonzero contributions to the integral come only from E_a (which can be taken as the lower limit of the integral), where $\sigma(E_{tr}) = \pi d^2$. Using an analytical approximation to the rate constant Eq. (1.27), one obtains[10] the approximate solution[11]

$$k(T) = \left(\frac{8k_BT}{\pi\mu}\right)^{1/2} \pi d^2 e^{-E_a/k_BT} = AT^{1/2}e^{-E_a/k_BT} \tag{1.59}$$

where $A = \left(\frac{8k_B\pi}{\mu}\right)^{1/2} d^2$. This formulation of $k(T)$ coincides with that of the kinetic theory of gases that expresses it as the product of the mean velocity of a particle $(8k_BT/\pi\mu)^{1/2}$ times the cross-sectional area of the particle (πd^2) associated with the fraction of effective collisions ($e^{-E_a/kT}$).

Equation 1.59 can be improved by multiplying A by a steric factor g. The factor g is a constant that takes into account the fact that molecules and related interactions do not, in general, have spherical symmetry and that, as a consequence, reactivity varies with the angle of collision. In this case, you can give g an angular dependence having a closed form that is simple, analytical, and integrable (for example, an ellipse as is often done for the study of liquids). The model can also take into account the fact that for molecules the collision involves other N degrees of freedom. In this case, if the interaction is expressed as $V(r, \xi_1, \xi_2, \xi_3, ..., \xi_N)$, the integral in Eq. 1.44 has the form

$$\int_a^\infty \mathrm{d}r \int_{a_1}^{b_1} \mathrm{d}\xi_1 \int_{a_2}^{b_2} \mathrm{d}\xi_2 ... \int_{a_N}^{b_N} \mathrm{d}\xi_N \frac{b}{r^2\left[1 - b^2/r^2 - V(r, \xi_1, \xi_2, \xi_3, ..., \xi_N)/E\right]^{1/2}},$$

where a_i and b_i are the turning points of the variable ξ_i.

1.4.2 *The Repulsive Coulomb Potential*

The Coulomb potential is a particular case of the family of the repulsive potentials $V(r) = Br^{-\delta}$, where δ is set equal to 1 (see Fig. 1.12). Typically, the part of the repulsive potential of a diatomic molecule is described by an index δ that varies between 9 and 15.

---

[10] $\int xe^{-x}\mathrm{d}x = -(e^{-x} + xe^{-x})$.

[11] $e^{-E_a/k_BT} + \frac{E_a}{k_BT}e^{-E_a/k_BT} \simeq e^{-E_a/k_BT}$ provided $1 >> E_a/k_BT$.

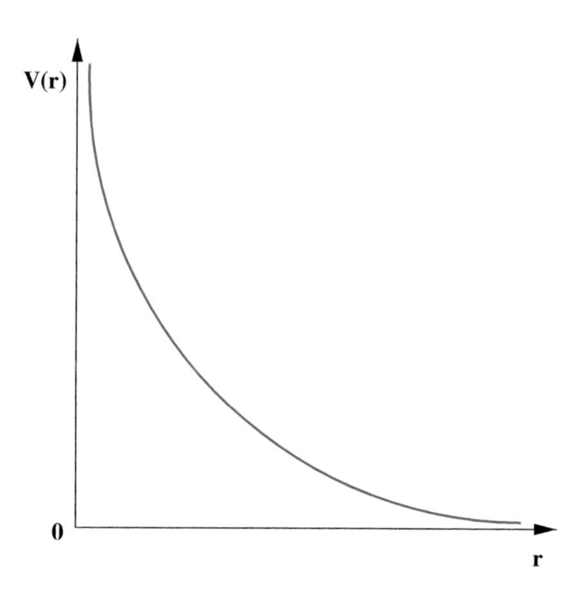

Fig. 1.12 Repulsive Coulomb potential defined as $V(r) = \frac{B}{r}$ with $B > 0$. For two electric charges $Z_1 e$ and $Z_2 e$, the interaction is $B = Z_1 Z_2 e^2/r$

On inserting the Coulomb potential, the integral of Eq. 1.44 becomes

$$\int_a^{\infty} r^{-2}\left(1 - \frac{b^2}{r^2} - \frac{B}{Er}\right)^{-1/2} \mathrm{d}r \tag{1.60}$$

and on putting $z = 1/r$, as previously done, the integral is transformed into

$$\int_0^{1/a} \left(1 - \frac{B}{E}z - b^2 z^2\right)^{-1/2} \mathrm{d}z \quad \text{with} \quad a = \frac{B}{2E} \pm \frac{1}{2}\left[\frac{B^2}{E^2} + 4b^2\right]^{1/2}. \tag{1.61}$$

The integral has now the following closed-form solution:[12]

$$\frac{1}{b}\left| arcsin \left[(2b^2 z + B/E)\left(\frac{B^2}{E^2 + 4b^2}\right)^{-1/2}\right]\right|_0^{1/a} \tag{1.62}$$

from which we obtain[13]

$$\theta = 2\, arcsin \left[1 + \frac{4b^2 E^2}{B^2}\right]^{-1/2}. \tag{1.63}$$

[12] $\int (1 - hx - cx^2)^{-1/2}\mathrm{d}x = -(c)^{-1/2}\, arcsin\left[-(cx + 2h)/(h^2 + 4c)^{-1/2}\right]$.
[13] $arcsin(x_1) \pm arcsin(x_2) = arcsin\left[x_1(1 - x_2^2)^{1/2} \pm x_2(1 - x_1^2)^{1/2}\right]$.

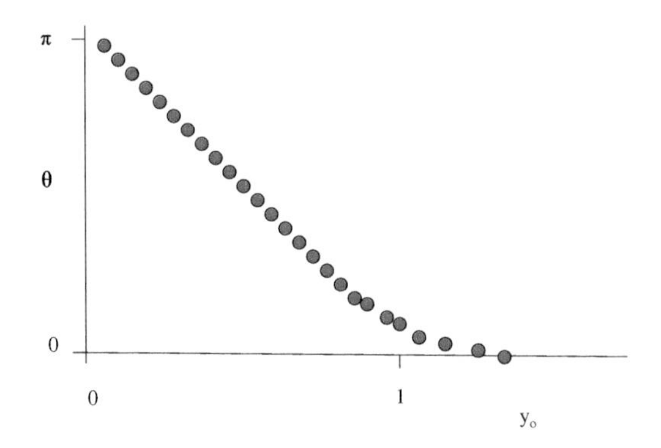

Fig. 1.13 Deflection angle as a function of the variable y_0 defined in Eq. (1.66) for a potential $V(r) = Br^{-12}$ for positive B

For the differential cross section, one has, using the previous relationship,

$$b = (B/2E)\cos\theta/2 \tag{1.64}$$

and then, according to the definition of cross section (see Eq. 1.53), we obtain the Rutherford formula:

$$\sigma(\theta, E_{tr}) = \left(\frac{B}{4E}\right)^2 \csc^4(\theta/2). \tag{1.65}$$

In the case of a generic central repulsive potential $V(r) = Br^{-\delta}$ with $\delta \neq 1$, no analytical solution to the integral can be worked out; hence, it is necessary to resort to a numerical quadrature. The calculation is simplified by introducing the reduced variables:

$$y = \frac{b}{r}, \quad y_a = \frac{b}{R_A}, \quad y_0 = b\left[\frac{E}{\delta B}\right]^{1/\delta}, \tag{1.66}$$

under which the deflection angle is given by

$$\theta = \pi - 2\int_0^{y_a}\left[1 - y^2 - \frac{1}{\delta}\left(\frac{y}{y_0}\right)^{\delta}\right]^{-1/2} \mathrm{d}y, \tag{1.67}$$

whose representation as a function of y_0 is given in Fig. 1.13.

1.4.3 Sutherland *and* Morse *attractive–repulsive potentials*

To rationalize most of the scattering features, however, one needs to add a long-range attractive tail to a short-range repulsive component. The simplest model potential of this type is the Sutherland one defined as $V(r) = \begin{cases} -\frac{q}{r^\gamma} & \text{if } r \geq a \\ \infty & \text{if } r < a \end{cases}$ where $q > 0$.

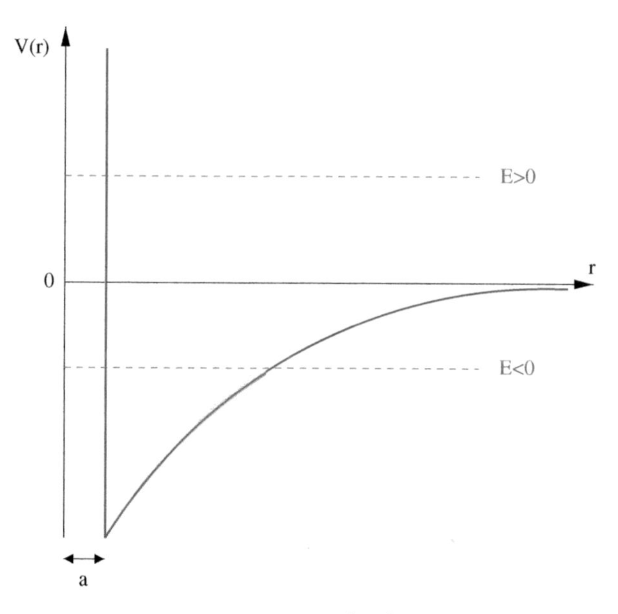

Fig. 1.14 Sutherland potential defined as $V(r) = \begin{cases} -\frac{q}{r^\gamma} & \text{if } r \geq a \\ \infty & \text{if } r < a \end{cases}$ with $q > 0$. For $E < 0$ (depending on the value of the impact parameter b), the particle can either be captured or not by the well, see Eq. (1.71)

The Sutherland potential is characterized by a repulsive rigid sphere behavior (infinitely repulsive at short distance) and a negative coulomb long-range one (see Fig. 1.14). Although the Sutherland model potential is quite rudimentary, it bears the interesting property of being attractive at long range and repulsive at short range.

For $\gamma = 1$ (attractive interaction of the Coulomb type), the deflection angle is

$$\theta = \pi - 2b \int_a^\infty r^{-2} \left[1 - \frac{b^2}{r^2} + \frac{q}{rE} \right]^{-1/2} dr. \tag{1.68}$$

Again letting $z = 1/r$ (and thus $dz = -\frac{1}{r^2}dr$), the integral is transformed into

$$\int_0^{1/a} \left[1 - b^2 z^2 + \frac{qz}{E} \right]^{-1/2} d\,z \tag{1.69}$$

having a solution similar to that of the Coulomb potential

$$-\frac{1}{b} \arcsin \left[\frac{-\left(-\frac{q^2}{E} z + 2b^2 z\right)}{\frac{q^2}{E} + 4b^2} \right] \tag{1.70}$$

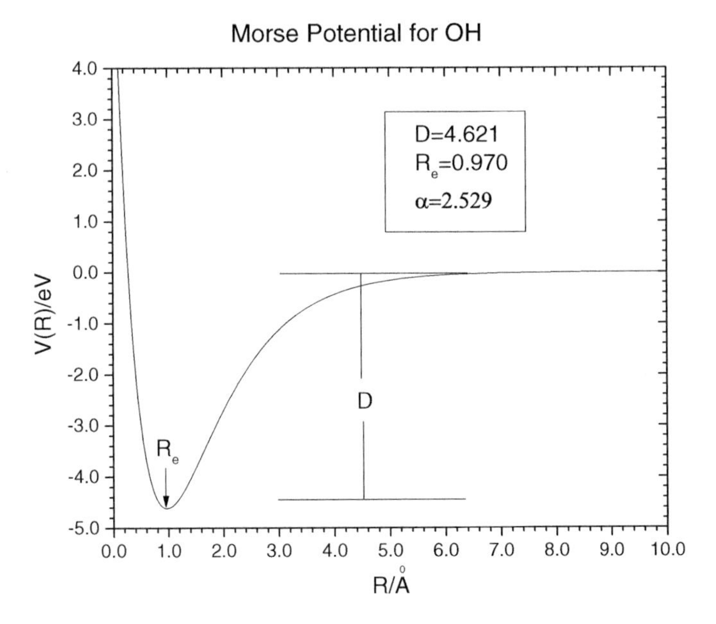

Fig. 1.15 The Morse potential (in eV) for the OH ($^2\Pi$) molecule plotted as a function of the internuclear distance given in Å. D, the dissociation energy, is 4.621 eV, R_e, the equilibrium distance, is 0.970 Å, and α, the exponential parameter, is 2.529 Å^{-1}

from which using the formula given in the footnote to the Coulomb potential to sum the inverse trigonometric functions, we have

$$\theta = \pi - 2\arcsin\left[\frac{1}{a}(b^2 + q/E)\right]^{1/2}. \tag{1.71}$$

In the case where $b^2 + q/E < 0$, there is no real solution.

A smoother and more realistic attractive–repulsive potential widely used for modeling diatomic molecules is the Morse one (see Fig. 1.15 where for illustrative purposes that of OH is considered[14]). The Morse potential is defined as

$$U(r) = D\left[e^{-2\alpha(r-r_e)} - 2e^{-\alpha(r-r_e)}\right] = D(n^2 - 2n) \tag{1.72}$$

with $n = e^{-\alpha(r-r_e)}$, the so-called bond order (BO) variable, being the building block of the interaction. The formulation of the two-body potential in terms of powers of the BO variables bears clear advantages to scattering calculations. It is, in fact, smooth and compact (it is made of two n terms that smoothly connect each other of which the first power characterizes the attractive part that naturally goes to zero

[14]The parameters for the OH molecule reported in the figure have been taken from *G. Herzberg*, Constant of Diatomic Molecules (Van Nostrand, 1978, New York).

as the internuclear distance r goes to infinity, while the second gets increasingly repulsive as the internuclear distance tends to zero). Furthermore, as we shall see in more detail later, it leads straightforwardly either to analytical or to easy to compute formulations of other diatomic properties, it is easy to generalize to higher powers of n and more bodies and it allows as well the formulation of simple continuity variables connecting different processes.[4]

1.4.4 The Scattering Lennard–Jones (6–12) potential

A popular formulation of the two-body potential in *scattering* is the LJ (6–12) potential (see Fig. 1.16). The model is particularly well suited for spherical nonpolar atoms. It has, in fact, a realistically repulsive short-range behavior (the choice of a 12th power in R to formulate the repulsive component is in large part due to the advantage of allowing analytical solutions) and mimics quite well the attractive van der Waals interaction at large radial distances. Its analytical form

$$V(r) = 4\epsilon\left[\left(\frac{\sigma}{r}\right)^{12} - \left(\frac{\sigma}{r}\right)^{6}\right] = \epsilon\left[\left(\frac{r_e}{r}\right)^{12} - 2\left(\frac{r_e}{r}\right)^{6}\right] \tag{1.73}$$

is given in terms of the dissociation energy ϵ, the equilibrium distance R_e, and the intercept of the potential with zero σ.

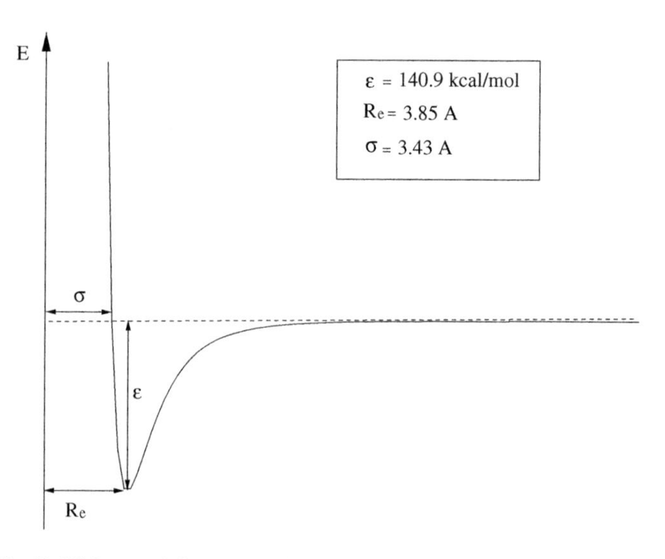

Fig. 1.16 The (6–12) Lennard–Jones potential for the HF molecule. Highlighted are the well depth ϵ that occurs for $R_e = 2^{1/6}\sigma$, the equilibrium distance R_e, and the parameter σ (the distance at which the potential crosses takes again the asymptotic limit value)

In the case of the LJ potential, the integral of Eq. 1.44 does not admit analytical solution. Accordingly, the calculation of θ is only possible using digital techniques. Anyway, it is instructive to compare the shape of this potential by plotting its value as a function of the impact parameter so as to highlight the effect of the angular momentum term on the effective potential $V^l(r)$ (see Eq. 1.41). The Hamiltonian of the system (see Eq. 1.33) can in fact be conveniently rewritten by expressing the angular velocity in terms of the two-body total angular momentum **L** (Eq. (1.40) from which $\dot{\theta} = l/\mu r^2$) leading to the following expression:

$$\mathcal{H} = \frac{1}{2}\mu\dot{r}^2 + \frac{l^2}{2\mu r^2} + V(r) = \frac{1}{2}\mu\dot{r}^2 + V^l(r). \tag{1.74}$$

The proper way to scale the Lennard–Jones potential is to divide the radial distances by the parameter σ ($r^* = r/\sigma$ and $b^* = b/\sigma$) and the energies by the parameter ϵ ($V^* = V/\epsilon$ and $E^* = E/\epsilon$). The scaled quantities are marked with an asterisk. In Fig. 1.17, the scaled effective Lennard–Jones (12–6) potential of a molecule is represented by different values of l as a function of r^*. From the figure, it is clearly seen that as the angular momentum increases, i.e., with increasing impact parameter, the centrifugal term in Eq. (1.74) becomes more significant so as to mask the presence of the potential well. Similarly, the *classical turning point* (the point of closest approach in the classical sense, as defined in Eq. 1.45) moves progressively toward greater distances. That is, for high values of the angular momentum, the angle of deflection is influenced only by the centrifugal part of the potential.

Once you have performed the numerical quadrature of the integral of Eq. 1.44, you can calculate, point by point, the values of θ as a function of the impact parameter at different collision energies (see Fig. 1.18). The figure shows that θ is positive at low impact parameters, while it is negative at medium and large ones. Yet, at low energy, an increase of b leads in the negative region to a sharp decrease of θ going down to quite large negative values (sometimes more negative than $-\pi$ (solid circles), whereas at higher energies such transition is smoother going through the formation of a shallow well with a small minimum (diamonds). The rational for such a behavior can be found by integrating related classical trajectories. As shown

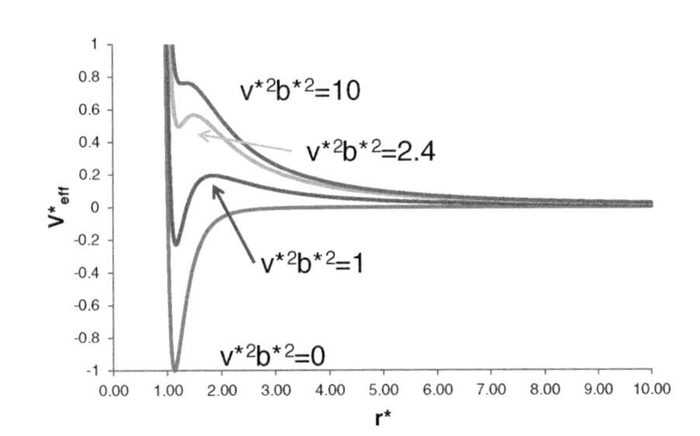

Fig. 1.17 Scaled Lennard–Jones effective potential (6–12): $V^*_{eff}(r) = V(r^*) + \frac{l^2}{2\mu r^{*2}} = V(r^*) + \frac{v^{*2}b^{*2}}{r^{*2}}$ for different values of $v^{*2}b^{*2}$

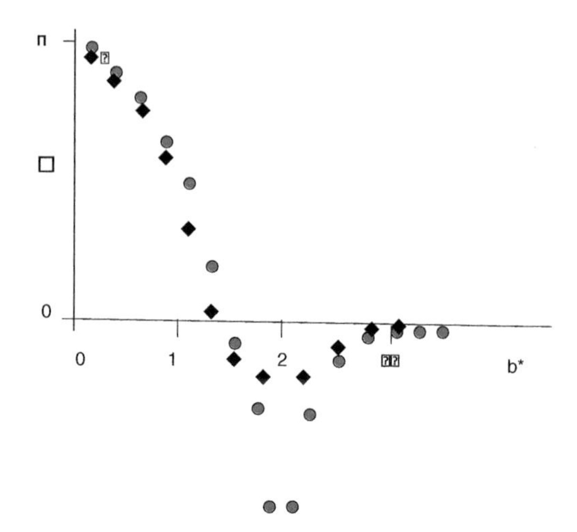

Fig. 1.18 Deflection angle θ (plotted as a function of the reduced impact parameter (b^*) for different values of the reduced energy (E^*)) computed on the Lennard–Jones (6–12) potential. Notice the difference between the diamond plot showing a rainbow feature (a small shallow minimum) and the solid circle plot showing an orbiting feature (a near singularity)

in Fig. 1.8, the low-energy sharp transition behavior can be ascribed to the orbiting capture of the trajectory. This is due to an almost even balance between attraction and escape tendency that leads to an exit in different (large negative) values of θ for small variations of b. These final values of θ may end up to coincide with the outcome of other orbiting (or non-orbiting) trajectories and will be the ground for rationalizing some interference effects in the next chapter. On the contrary in the higher energy regime, the trajectories associated with the solid circles show that there is not an orbiting capture. There is instead a limiting deflection angle leading to a small negative minimum that is usually called "rainbow" and offers a rationale for some interference effects that will be commented later. As b increases (see Fig. 1.17), very large centrifugal (repulsive) contributions almost entirely erase the potential well and make the deflection angle tend to zero.

1.5 Problems

1.5.1 Qualitative Problems

1. **Trajectories-Fixed Energy**: Without performing any calculations, describe the qualitative behavior for the classical trajectories for a fixed energy and varying the impact parameter from 0 to very large values. Provide a separate description for each of the four given potentials.
2. **Deflection Angle-Fixed Energy**: Without performing any calculations, describe the qualitative behavior for the deflection angle for a fixed energy and varying the impact parameter from 0 to very large values. Provide a separate description for each of the four given potentials.

3. **Trajectories-Fixed Impact Parameter**: Without performing any calculations, describe the qualitative behavior for the classical trajectories for a fixed impact parameter and varying the energy from 0 to very large values. Provide a separate description for each of the four given potentials.
4. **Deflection Angle-Fixed Impact Parameter**: Without performing any calculations, describe the qualitative behavior for the deflection angle for a fixed impact parameter and varying the energy from 0 to very large values. Provide a separate description for each of the four given potentials.

1.5.2 Quantitative Problems

1. **Potentials**: The Lennard–Jones and the Morse potential qualitatively look similar. The Lennard–Jones potential has two parameters (ϵ, r_e) or (ϵ, σ). How are r_e and σ related if the two forms of the potential are identical? The Morse potential has three parameters (ϵ, r_e, and β). Derive an expression for β to make the Morse potential have the same identical well depths ϵ, equilibrium positions r_e, and the value of r where they cross zero. Plot both potentials on the same graph and explain the difference that you see.

2. **Numerical Integration**: Use the midpoint integration rule to integrate the following integrals: $\int_0^2 x^2\, dx$, $\int_0^\infty e^{(-3r)}\, dr$, $\int_\sigma^\infty V_{LJ}(r)\, dr$, and $\int_\sigma^\infty V_{Morse}(r)\, dr$. For the Lennard–Jones and Morse potential, use $\epsilon = 140.9$ kcal/mole, and $r_e = 3.85$Å and choose β in the Morse potential so it crosses zero energy at the same distance as the Lennard–Jones potential. Which potential form might be better suited for scattering at very low energies?

3. **Trajectories**: Write a program to calculate trajectories for central field potentials. When the potential and the impact parameter are both zero, does your program produce correct results? Explain the trajectories you produce when the potential is zero but the impact parameter $b > 0$. Now use the Lennard–Jones potential with $\epsilon = 140.9$ kcal/mole and $r_e = 3.85$Å. Using your trajectory code to create a table of deflection angles for 11 energies in the range $E = [20, 120]$ kcal/mole and 11 impact parameters in the range $b = [0, 10]$Å. Justify your results. For an impact parameter of $b = 3$Å , find the value of the energy where an orbiting trajectory occurs.

4. **Deflection Angle**: Write a program to calculate the classical deflection angle. Use the same potential, scattering energies, and impact parameters as provided in the previous problem. Do these angles correspond to the deflection angles from your classical trajectory problem? Reproduce a plot similar to Fig. 1.7 that clearly shows rainbow scattering and orbiting behavior.

5. **Deflection Angle**: Use the same potential as used in the previous two problems to calculate classical action and the delay time. Provide a physical interpretation for these two quantities.

6. **Cross Section**: Use the same potential as used in the previous three problems to calculate classical differential cross section and total cross section. Provide a physical interpretation for these two quantities.

7. **Favorite Diatomic Molecule**: Pick your favorite diatomic molecule, then use the literature to find the well depth and the equilibrium position. Compare deflection angles, differential cross section, and total elastic cross section using both the Lennard–Jones potential and the Morse potential (with β constrained to have the same value of σ as the Lennard–Jones potential). Compare with experimental results if possible. Can you get a better agreement with the experimental results by adjusting β in the Morse potential?

Chapter 2
The Quantum Approach to the Two-Body Problem

This chapter, after considering the drawbacks of a classical mechanics treatment of the deflection angle in two-body collisions, tackles the problem of treating such collisions using a quantum approach to both bound states and elastic scattering. Some basic interaction models are then considered in order to provide support to some fundamental relationships between the shape of the interaction and the formulation of bound states and scattering quantities. Basic techniques often adopted for numerically integrating the Schrödinger equation are finally illustrated.

2.1 Quantum Mechanics and Bound States

2.1.1 The Limits of the Classical Mechanics Approach

So far, we have treated atomic and molecular processes as collisions of pointwise particles using the concepts of classical mechanics. In classical mechanics, each particle is assigned a well-defined position (say $\mathbf{x}$ in one dimension) and a momentum (say $\mathbf{p}_x = \mathrm{m}\mathbf{v}_x = \mathrm{m}d\mathbf{x}/\mathrm{dt}$) at each value of time t. This assumption turns out not to be valid in some cases (like in the case of light that in *Newton*'s view is treated as made of particles, whereas in the *Huygens*'s view is treated as made of waves propagating on an ether (like the waves generated by a stone thrown on water)). While ether was proven not to exist by some sophisticated experiments wave-like behaviors (such as diffraction patterns) were found to be typical of small particles (like photons, electrons, etc.). Thanks to *Born*, the behaviour of elementary particles was identified to be of probabilistic nature, that in one dimension can be associated with a wave function, say $\psi(x, t)$, representing the probability amplitude allowing the evaluation of related physical observables. In the *Dirac* notation, the function ψ can be written as a vector $|\psi\rangle$ (more details about vectors, matrices, and vector spaces

A. Laganà and G. A. Parker (eds.), *Chemical Reactions*, Theoretical Chemistry and Computational Modelling, https://doi.org/10.1007/978-3-319-62356-6_2

are given in Appendix A1). Based on an empirical assumption, the wave equation for light is written in one dimension (1D) as

$$\psi(x, t) = C_x \cdot e^{i\alpha_x} \tag{2.1}$$

where C_x is a normalization factor, α_x is the so-called phase factor $\alpha_x = 2\pi(x/\lambda_x - \nu_x t)$ with λ_x being the wavelength and ν_x the frequency given by the velocity of light c divided by λ_x.

The fundamental postulates of quantum mechanics are the discretization of energy (Plank)

$$E = h\nu \tag{2.2}$$

(a quantization recently extended to masses (gravitons) and time (chronons)) and the energy–mass relationship (Einstein)

$$E = mc^2. \tag{2.3}$$

By comparing the two expressions, one obtains $mc = h/\lambda$ that was generalized to all particles moving at υ smaller than the speed of light (De Broglie). By embodying the above postulates into Eq. 2.1 one obtains

$$\psi(x, t) = C_x \cdot e^{i(xp_x - Et)/\hbar}. \tag{2.4}$$

When Eq. 2.4 is differentiated by time (t), one obtains the 1D time-dependent Schrödinger equation

$$i\hbar\frac{\partial\psi}{\partial t} = E\psi \tag{2.5}$$

that allows us to define the energy operator $\hat{E}$ (marked by the "hat")

$$\hat{E} = i\hbar\frac{\partial}{\partial t} \tag{2.6}$$

delivering energy from the probability amplitude function ψ. When Eq. 2.4 is differentiated by space (x) one obtains

$$-i\hbar\frac{\partial\psi}{\partial x} = |p_x\rangle\psi \tag{2.7}$$

that allows us to define the momentum operator $\hat{p_x}$

$$\hat{p_x} = -i\hbar\frac{\partial}{\partial x}. \tag{2.8}$$

The different operators do not necessarily commute. Given two operators (say $\hat{A}$ and $\hat{B}$), their property of commuting is checked through the commutator

$$\left[\hat{A}, \hat{B}\right] = \hat{A}\hat{B} - \hat{B}\hat{A} \tag{2.9}$$

When two operators do not have common eigenvectors and their associated observables are not exactly determined at the same time their commutator is not a zero operator, $\hat{O}$, and they do not commute. It can be shown that this is the case of the position and momentum operators ($\hat{x}$ and $\hat{p_x}$) for which $\left[\hat{x}, \hat{p_x}\right] = i\hbar \neq 0$ (similarly for $\hat{t}$ and $\hat{E}$ one obtains $\left[\hat{t}, \hat{E}\right] = -i\hbar$). This means that these pairs of variables cannot be simultaneously determined. Formally, this is expressed by the uncertainty principle (Heisenberg relationship)[1]:

$$\Delta x \Delta p_x \geq \frac{\hbar}{2}. \tag{2.10}$$

To relate the uncertainty of b and θ in our choice of coordinates, one has to recall that the uncertainty relationship that has to be applied is $\Delta b \Delta p_z \simeq \hbar$. Because the z-component of the momentum p_z transferred during the collision is $p_z = \mu v \sin\theta \simeq \mu v \theta$, we get immediately

$$\Delta b \Delta \theta \simeq \hbar/\mu v \tag{2.11}$$

or

$$\mu v \Delta b \Delta \theta = \Delta L \Delta \theta \simeq \hbar. \tag{2.12}$$

Let us consider now the singularities associated with the classical formulation (see Eq. 1.54) of the differential cross section at both $\theta = 0$ and $\mathrm{d}\theta/\mathrm{d}b = 0$ as is the case shown in Fig. 2.1 for the Lennard-Jones, LJ, potential (see the end of the previous chapter). In the figure, the value of the deflection angle θ corresponding to a minimum (angle of rainbow) leads to a singularity corresponding in the quantum treatment to a broad maximum with superimposed a highly oscillating structure.

The fact that the singularity of the classical result is smoothed in the quantum solution (see Eq. 2.79) can be traced back to the uncertainty principle of Eq. 2.12 that does not allow θ and b to be simultaneously defined. This inability of a pure classical mechanics approach to embody such condition can be regained in semiclassical treatments by taking into account the interference effects between waves associated with different classical paths (trajectories) though leading to the same value of θ. This result provides the theoretical basis for defining the abovementioned wave function ψ as a scattering physical distribution in space (obviously with different probabilities in different regions) of the system. Given a precise value L (or equivalently of

[1] The original heuristic argument was given by Heisenberg in 1927. The formal statement was proved by Earl Hesse Kennard and Hermann Weyl.

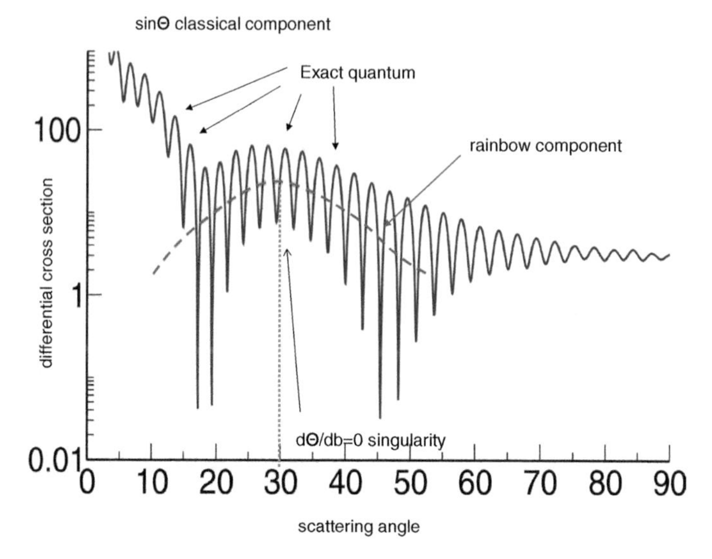

Fig. 2.1 Classical differential cross section and its components for a potential Lennard-Jones (6–12) plotted as a function of the scattering angle Θ ($\Theta = abs(\pi - \theta)$) compared with corresponding quantum value (solid highly oscillating line)

the impact parameter b) of the total angular momentum L, the value of the deflection angle θ should be completely uncertain.

2.1.2 *The 3D Quantum Problem and Its Decomposition*

The extension of the 1D Eq. 2.7 in x to three dimensions calls for the inclusion of the corresponding terms in y and z. The three-dimensional quantum Hamilton operator $\hat{H}$ is

$$\hat{H} = \hat{T} + \hat{V} = -\frac{\hbar^2}{2\mu}\nabla^2 + \hat{V} \tag{2.13}$$

that includes the Laplacian ∇^2 (see Appendix A1) and the potential $\hat{V}$.

The idea that for every molecular system there exists a stable energy state with a well-defined energy (called the ground state) associated with the equilibrium geometry of the system is largely accepted and is supported by experimental evidence. It is also generally accepted that there are stable (or metastable) energies which are higher than the ground state in which the system can exist for long intervals of time. It should be emphasized that in reality only the ground state will be indefinitely stable. All higher energy states will eventually decay. However, in the present text, we will not use relativistic quantum mechanics or quantum electrodynamics to treat these decays. The fact that the energy has an exact discrete value requires us (via the Heisenberg uncertainty principle) to give a description of the system which is distributed in space with a probability of finding the system in the corresponding configuration proportional to the square modulus of the wave function Ψ. The properties of Ψ are determined by applying the appropriate operators (corresponding quantum operators of classical variables). In the Cartesian coordinate system, these operators take simple forms ($\hat{p}_x$, the momentum operator along x, becomes $-i\hbar(\partial/\partial x)$ for which the corresponding kinetic term of the Hamiltonian becomes $-\hbar^2/2\mu(\partial^2/\partial x^2)$ and in the case of a system of two particles that, as already shown, is isomorphic to the problem of a particle of mass μ subject to the central potential $V(r)$ is described by the time- dependent Schrödinger equation

$$\hat{H}\Psi(\mathbf{r},t) = i\hbar\frac{\partial}{\partial t}\Psi(\mathbf{r},t). \tag{2.14}$$

that can be integrated as a first-order equation in time t. As you see, we have a vector coordinate in $\mathbf{r}$ and, correspondingly, a Hamiltonian $\hat{H}$ containing the conjugated momentum operator $\hat{p}_{\mathbf{r}}$ in addition to the potential $V(r)$. The measurable quantity is the expectation of the momentum operator $\langle\Psi\left|\hat{p}_{\mathbf{r}}\right|\Psi\rangle$ where Ψ is the probability amplitude whose square is the probability density. In general, for closed systems (whose Hamiltonian is not time dependent) it is preferred to use separation of variables to factorize the time dependence as follows:

$$\Psi(\mathbf{r},t) = \Psi(\mathbf{r})\chi(t). \tag{2.15}$$

Thanks to the separation of the time variable in Eq. 2.15 one can write

$$i\hbar\frac{\partial}{\partial t}\Psi(\mathbf{r},t) = E\Psi(\mathbf{r},t) \tag{2.16}$$

or equivalently

$$i\hbar\frac{\partial}{\partial t}\chi(t) = E\chi(t) \tag{2.17}$$

from which the solution for the time-dependent component $\chi(t) = e^{-iEt/\hbar}$ can be obtained. At the same time, the stationary (time independent) Schrödinger equation reads

$$\left[-\frac{\hbar^2}{2\mu}\nabla_{\mathbf{r}}^2 + V(r)\right]\Psi(\mathbf{r}) = E\Psi(\mathbf{r}) \tag{2.18}$$

that can also be written in the synthetic form

$$\left(\nabla_{\mathbf{r}}^2 - U(r) + k^2\right)\Psi(\mathbf{r}) = 0 \tag{2.19}$$

where $\nabla_{\mathbf{r}}^2$ is the sum of the Cartesian Laplacian components in x, y and z and is proportional to the kinetic operator, $U(r)$ is the potential energy $V(r)$ scaled by the mass μ

$$U(r) = \frac{2\mu V(r)}{\hbar^2}$$

and k is the modulus of the wave vector of the incident particle

$$k = \left(\frac{2\mu}{\hbar^2}E\right)^{1/2} = \frac{2\pi\mu v}{h}. \quad (2.20)$$

There are different techniques to solve the stationary Schrödinger equation.[2] The Schrödinger equation (2.19) is a second-order partial differential equation (PDE) of elliptic type. Its integration is carried out, usually, either using numerical techniques or separating variables. In this regard, the Laplacian operator $\nabla_{\mathbf{r}}^2$ is expressed in spherical polar coordinates[3] (r, ϑ, ψ) as illustrated in the LHS scheme of Fig. 1.5

$$\nabla_{\mathbf{r}}^2 = \frac{1}{r^2}\frac{\partial}{\partial r}\left(r^2\frac{\partial}{\partial r}\right) + \frac{1}{r^2 \sin\vartheta}\frac{\partial}{\partial\vartheta}\left(\sin\vartheta\frac{\partial}{\partial\vartheta}\right) + \frac{1}{r^2 \sin\vartheta}\frac{\partial^2}{\partial\psi^2} \quad (2.21)$$

whose component in r can also be written (see Appendix A2) as

$$\frac{\partial^2}{\partial r^2} + \frac{2}{r}\frac{\partial}{\partial r} = \frac{1}{r}\frac{d^2}{dr^2}r \quad (2.22)$$

Equation 2.22 is particularly useful in different situations as we will see later in this text.

The function $\Psi(\mathbf{r})$ is, in turn, expressed as a product of a radial term $R(r)$ and an angular term $Y(\vartheta, \psi)$

$$\Psi(\mathbf{r}) = R(r)Y(\vartheta, \psi). \quad (2.23)$$

By substituting this product function in the Schrödinger equation (2.19) and separating the radial from the angular terms, we have:

$$\frac{1}{R(r)}\frac{\partial}{\partial r}\left(r^2\frac{\partial R(r)}{\partial r}\right) + r^2\left(k^2 - U(r)\right) = \quad (2.24)$$
$$-\frac{1}{Y(\vartheta, \psi)}\left[\frac{1}{\sin\vartheta}\frac{\partial}{\partial\vartheta}\left(\sin\vartheta\frac{\partial Y(\vartheta, \psi)}{\partial\vartheta}\right) + \frac{1}{\sin\vartheta}\frac{\partial^2 Y(\vartheta, \psi)}{\partial\psi^2}\right]$$

that is based on the symmetry relationships of the system. This is due to the isotropy and uniformity of the space because the angular momentum $\mathbf{L}^2$ and its projection

[2] The integration of the equation of the time-dependent Schrödinger uses different techniques that will be discussed later for the case of the atoms reacting with diatoms. For a discussion details see, for example, Ref. [5].

[3] In mathematical texts one usually see the ϑ as azimuthal angle and ψ as the polar angle. However, in almost all physics or chemistry texts ϑ is polar angle and ψ is the azimuthal angle.

L_z on a space-fixed coordinate system are conserved[4] with value $[l(l+1)]$ and m_l Accordingly, the separation constant of the LHS and RHS terms of Eq. 2.24 is set equal to the discrete variable $[l(l+1)]$. For the LHS, we will have, therefore,

$$\frac{1}{r^2}\frac{\mathrm{d}}{\mathrm{d}r}\left(r^2\frac{\mathrm{d}\,R(r)}{\mathrm{d}\,r}\right)+\left[k^2-U(r)-\frac{l(l+1)}{r^2}\right]R(r)=0 \tag{2.25}$$

as we noted before in Eq. 2.22, this equation can be written as

$$\frac{1}{r}\frac{\mathrm{d}^2}{\mathrm{d}r^2}\left[rR(r)\right]+\left[k^2-U(r)-\frac{l(l+1)}{r^2}\right]R(r)=0. \tag{2.26}$$

This latter expression suggests that we introduce the function $R(r)=\eta_l(r)/r$ (we have indicated explicitly with the subscript l the fact that we have a radial function for every l). In this way, we obtain

$$\left[\frac{\mathrm{d}^2}{\mathrm{d}r^2}+k^2-U^l(r)\right]\eta_l(r)=0 \tag{2.27}$$

with an effective potential

$$U^l(r)=U(r)+\frac{l(l+1)}{r^2}$$

while for the RHS terms

$$\frac{1}{\sin\vartheta}\frac{\partial}{\partial\vartheta}\left(\sin\vartheta\frac{\partial}{\partial\vartheta}+\frac{1}{\sin\vartheta}\frac{\mathrm{d}^2}{\mathrm{d}\psi^2}\right)Y(\vartheta,\psi)+l(l+1)Y(\vartheta,\psi)=0. \tag{2.28}$$

As such we have decomposed the original problem, defined by Eq. (2.18), into a subproblem for the radial equation (2.27) and a subproblem for the angular equation (2.28). The angular equation is the equation of the spherical harmonics $Y_{lm}(\vartheta,\psi)$, which is a very important relation in quantum treatments. Here, it is worth remembering that the functions $Y_{lm}(\vartheta,\psi)$ can be expressed as

$$Y_{lm}(\vartheta,\psi)=\sqrt{\frac{(2l+1)}{4\pi}\frac{(l-m)!!}{(l+m)!!}}P_l^m(\cos\vartheta)e^{im\psi} \tag{2.29}$$

in terms of associated Legendre polynomials ($P_l^m(\cos\vartheta)$) and exponentials $e^{im\psi}$ and that the spherical harmonics have a fundamental importance for the algebra of angular momenta. The spherical harmonics belong to the family of those functions that are called special functions, because their properties are usually linked to the features

[4]If there was an external magnetic field then it will break the isotropic symmetry and then L^2 is no longer conserved.

of differential equations and are based on higher level algebraic treatments. Among the simplest (and of widespread use) are the special functions Gamma, $\Gamma(z)$, (a generalization of the factorial) defined by the integral

$$\Gamma(z) = \int_0^\infty t^{z-1} e^{-t} dt \tag{2.30}$$

although most often calculated from the relationship

$$\Gamma(z+1) = z\Gamma(z) \tag{2.31}$$

The so-called mother of all special functions is the hypergeometric function which we will discuss shortly.

2.1.3 *The Harmonic Oscillator*

The radial part of the solution, in general, is obtained by using numerical techniques except in some cases where for specific model problems the solution can be obtained in closed form. One such case, which is commonly used, is that of a particle moving under the influence of a linear restoring force $F(x) = -kx$ (i.e., subject to a potential $V(x) = \frac{1}{2}kx^2$) is the harmonic oscillator (HO). The harmonic oscillator is used to study the quantum nature of light (photons) and is also important in chemical physics, because it is often used as an initial approximation in the calculation of the vibrational frequencies of polyatomic molecules and crystal lattices. Such a potential ($V(x) \rightarrow \infty$ for $x \rightarrow \pm\infty$) admits only bound solutions (no dispersion) and represents the extreme case of total energy E always lower than the potential asymptote.

From the classical treatment, we know that a particle of mass m which is subject to this potential[5] has an angular frequency ω (or angular displacement per unit time)

$$\omega = \sqrt{k/m} = 2\pi\nu$$

where ν is the frequency of rotation usually measured in hertz.
Then expressing the potential as $V(x) = (m/2)w^2x^2$ the Schrödinger equation describing the one-dimensional system is

$$\left(-\frac{\hbar^2}{2m}\frac{\mathrm{d}^2}{\mathrm{d}x^2} + \frac{m\omega^2 x^2}{2}\right)\phi(x) = E\phi(x) \tag{2.32}$$

[5]The discussion attains of course also to the relative motion of two particles of mass m_1 and m_2 interacting with this potential because, as already pointed earlier, the problem of the motion of two particles is isomorphous with that of a single particle having a mass equal to the reduced mass ($\mu = m_1m_2/(m_1 + m_2)$) of the two particle one.

At this point, it is useful to introduce the dimensionless or reduced variables[6]

$$\xi = \frac{x}{\sigma_0} \text{ with } \sigma_0 = \sqrt{\frac{\hbar\omega}{m}}$$

so that Eq. (2.32) takes the form

$$\frac{1}{2}\left(-\frac{d^2}{d\xi^2} + \xi^2\right)\phi(\xi) = \epsilon\phi(\xi) \text{ with } \begin{cases} \phi(\xi) = \phi(x/\sigma_0) \\ \epsilon = E/\hbar\omega \end{cases} \quad (2.33)$$

where the eigenvalues ϵ are also dimensionless quantities that give the energy of the oscillator in multiples of Planck's energy quanta ($\hbar\omega$). The solutions of the Eq. (2.33) can be written in the form[7]

$$\phi_n(\xi) = N_n H_n(\xi) e^{-\xi^2/2} \quad (2.34)$$

where the Hermite polynomials $H_n(\xi)$ are orthogonal polynomials of degree n in ξ and the factor N_n is a constant of normalization. This can be obtained by imposing the usual normalization condition for the functions $\phi_n(\xi)$, or $\int_{-\infty}^{\infty} \phi_n^2(\xi) d\,\xi = 1$ getting

$$N_n = \left(\sqrt{\pi} n^2 n!\right)^{-1/2}. \quad (2.35)$$

In Fig. 2.2, we report the Hermite polynomials and the corresponding HO eigenfunctions for the first five values of n.

The extension of this discussion to the case of the three-dimensional HO is immediate. The potential $V(\mathbf{r}) = (m/2)\omega^2(x^2 + y^2 + z^2)$ allows, in fact, a separation of variables that leads to the solution of three equations of the type (2.33), one for each coordinate. The eigenvalues of the 3D oscillator will, therefore, take the form

$$E_n = \hbar\omega(n_1 + n_2 + n_3 + 3/2) \equiv \hbar\omega(n + 3/2) \quad (2.36)$$

and each level is degenerate $(n + 1)(n + 2)/2$ times. The eigenfunction results from the product of three functions of the type given in (2.34) (for a graphical representation of some of these functions see [6]).

Note, however, that in the case of the harmonic diatomic oscillator in r of Eq. 2.27, the r variable spans the range $(0, \infty)$ whereas the variable x has the range $(-\infty, \infty)$. Accordingly, the radial harmonic potential $U(r)$ is not symmetric and related η functions eigensolutions of Eq. 2.27 result in either even or odd Hermite polynomials and eigenfunctions, i.e., $\eta_n(-\xi) = (-1)^n \eta_n(\xi)$. This means that the potential is no longer symmetric in ξ and the HO wave functions no longer have proper symmetry.

---

[6]Reduced variables are also used for other potential models like the Lennard-Jones.

[7]The form of this function can be obtained by the same considerations made in the previous chapter. In fact, for $\xi \to \infty$ Eq. 2.33 turns into $d^2\phi/d\xi^2 = \xi^2\phi$ the solution of which is the function $e^{-\xi^2/2}$.

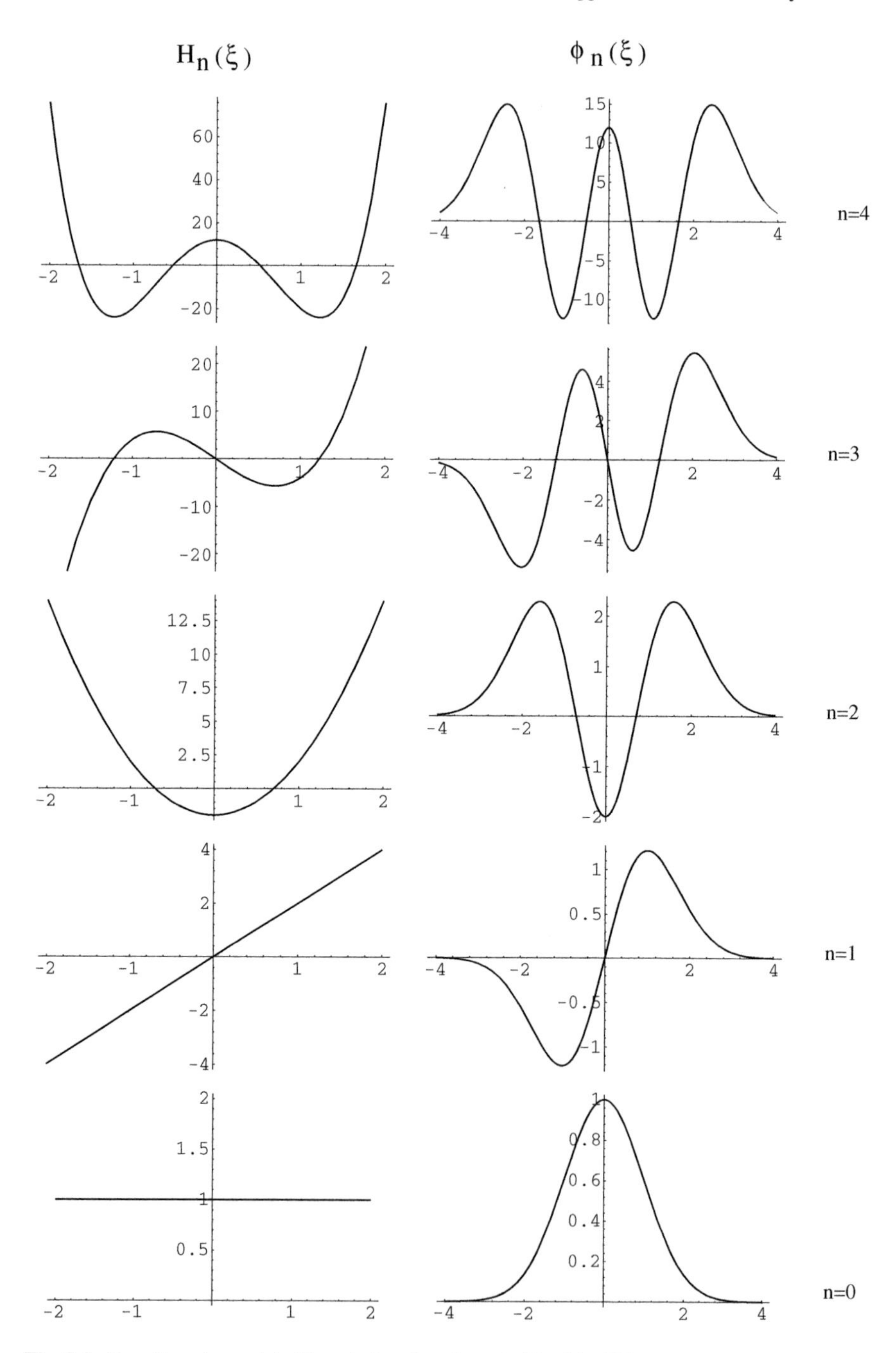

Fig. 2.2 Hermite polynomials H_n and eigenfunctions $\phi_n(\xi)$ of the HO

To get the correct eigenfunctions, we need to make a linear combination of odd and even terms

$$\eta_n(\xi) = a\eta_{2n}(\xi) + b\eta_{2n+1}(\xi) \tag{2.37}$$

where $a^2 + b^2 = 1$ and the radial wave function must be zero at $r = 0$ with $a\eta_{2n}(-r_e) + b\eta_{2n+1}(-r_e) = 0$ giving the second condition to determine the coefficients a and b. In fact, if $r_e = 0$ then $a = 0$ and $b = 1$. The linear combination of the even and odd eigenfunctions is, therefore, as follows:

$$\eta_n(\xi) = A_n N_n H_n(\xi)e^{-\xi^2/2} + B_n N_{n+1} H_{n+1}(n+1)(\xi)e^{-\xi^2/2} \tag{2.38}$$

giving

$$E_n = A_n\hbar\omega\,[A_n(n+3/2) + B_n(n+5/2)] \tag{2.39}$$

where A_n and B_n are chosen to make $\eta_n(\xi_{r=r_e}) = 0$ and normalized. As an illustration, let $r_e = 0$ then we have half of the HO with the condition that at $r = 0$ the wave function is zero for odd parity. This only occurs for odd values of n and therefore

$$\eta_n(\xi) = (2n+1)N_{2n+1}H_{2n+1}(\xi)e^{-\xi^2/2} \tag{2.40}$$

giving energy eigenvalues, $E_n = \hbar\omega(2n + 5/2)$. If one includes angular momentum, the exact eigenenergies are given by the expression $E_n = \hbar\omega(2k + l + 3/2)$ with l being the angular momentum quantum number. This is equivalent to Eq. 2.36 with a replacement of l by an odd integer. For an even and an odd value of l, one has respectively an even and an odd parity defined like $p = (-1)^l$.

One should emphasize here again that in the above description of the HO we used the variable ξ with range $(-\infty, \infty)$ with the potential being symmetric about $\xi = 0$.

2.2 Quantum Elastic Scattering

2.2.1 *The Coulomb Potentials and the Hydrogen Atom*

The Coulomb potential:

$$V(r) = \pm q/r \quad q = \text{positive constant}$$

is even more emblematic and important. For example, the solution for the attractive $V(r) = -q/r = -Ze^2/r$ potential of Z protons and one electron can be obtained in closed form. For both the hydrogen atom (one proton) and for a high lying Rydberg state of a many-electron atom, $Z = 1$ and the interaction is attractive[8]

We have already seen that the Schrödinger equation can be rewritten in spherical polar coordinates and the wave function can be formulated as the product of a radial function and an angular function[9]

$$\Psi(\mathbf{r}) = R(r)Y(\vartheta, \psi). \tag{2.41}$$

In this way, the Schrödinger equation can be decomposed into a radial equation containing the Coulombic potential

$$\hat{H}_r R(r) = \frac{\mathrm{d}^2 R(r)}{\mathrm{d}r^2} + \frac{2}{r}\frac{\mathrm{d}R(r)}{\mathrm{d}r} - \frac{l(l+1)}{r^2}R(r) + \frac{2\mu}{\hbar^2}\left(E + \frac{q}{r}\right)R(r) = 0 \tag{2.42}$$

and an angular one that is

$$\hat{H}_{\vartheta,\psi}Y(\vartheta, \psi) = \left[\frac{1}{r^2 \sin\vartheta}\frac{\mathrm{d}}{\mathrm{d}\vartheta}\left(\sin\vartheta\frac{\mathrm{d}}{\mathrm{d}\vartheta}\right) + \frac{1}{r^2 \sin\vartheta}\frac{\mathrm{d}^2}{\mathrm{d}\psi^2}\right]Y(\vartheta, \psi) = 0. \tag{2.43}$$

In atomic units[10] with $Z = 1$ and the reduced mass, μ approximated as m_e the radial equation (2.42) reads

$$\frac{\mathrm{d}^2 R(r)}{\mathrm{d}r^2} + \frac{2}{r}\frac{\mathrm{d}R(r)}{\mathrm{d}r} - \frac{l(l+1)}{r^2}R(r) + 2\left(E + \frac{1}{r}\right)R(r) = 0. \tag{2.44}$$

so by introducing the two variables (see [7])

$$n = \frac{1}{\sqrt{-2E}} \quad \text{and} \quad \rho = \frac{2r}{n} \tag{2.45}$$

it becomes

$$\frac{\mathrm{d}^2 R(\rho)}{\mathrm{d}\rho^2} + \frac{2}{\rho}\frac{\mathrm{d}R(\rho)}{\mathrm{d}\rho} + \left[-\frac{1}{4} + \frac{n}{\rho} - \frac{l(l+1)}{\rho^2}\right]R(\rho) = 0. \tag{2.46}$$

[8] $V(r) = -Ze^2/r$ in cgs unit and $V(r) = -Z^2/(4\pi\epsilon_0)r$ (with ϵ_0 being the permittivity in a vacuum) in S.I. unit. Still, $V(r) = -Z^2/(4\pi\epsilon_0)r = -(Z/r)\beta\hbar c$ where β is the fine structure constant ($\beta \approx 1/137$) and c is the speed of light in vacuum.

[9] For the motion in a central field, it is always possible to separate variables adopting a system of spherical polar coordinates. In the particular case of the Coulomb potential, separation of variables can also be carried out in parabolic coordinates which are useful for some applications, see [7, 8].

[10] The Energy unit is in fact equal to $m_e e^4/\hbar^2$. It corresponds to $\approx 27.211 eV \approx 627,509\, kcal/mol$ and is indicated with E_h. In atomic unit is $e = m_e = \hbar = 1$.

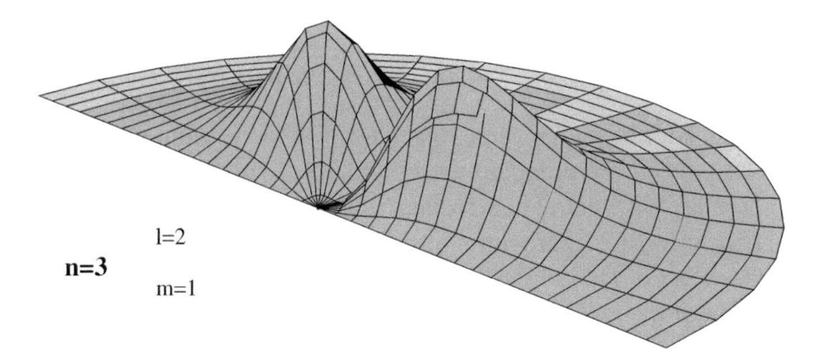

Fig. 2.3 The square of the absolute value, $|\psi_{nlm}(r, \vartheta, \psi)|^2$, three-dimensional wave function of the hydrogen atom for $n = 3$, $l = 2$ and $m = 1$

When the total energy E of the system (see Ref. 2.42) is negative (i.e., when E is lower than the asymptotic value of the potential energy ($V(\infty)$) that is taken as the energy zero ($E < 0$)), the radial function solution of (2.46) can be formulated as[11]

$$R(\rho) = \rho^l e^{-\rho/2} w(\rho). \tag{2.47}$$

Substituting (2.47) into Eq. (2.46), we obtain for $w(\rho)$ radial equation

$$\rho \frac{\mathrm{d}^2 w(\rho)}{\mathrm{d}\rho^2} + (2l + 2 - \rho)\frac{\mathrm{d}w(\rho)}{\mathrm{d}\rho} + (n - l - 1)w(\rho) = 0 \tag{2.48}$$

whose solution is the confluent hypergeometric function [9]

$$w(\rho) = F(-n + l + 1, 2l + 2; \rho). \tag{2.49}$$

The radial wave function $R(\rho)$ tends to zero both for $\rho \to 0$ and $\rho \to \infty$ (for $E < 0$), and therefore the function $w(\rho)$ must be finite for $\rho = 0$ and tends to zero faster than ρ^{-l} for large ρ. This requires that $w(\rho)$ is a finite polynomial. This happens only when $-n + l + 1$, the first parameter of $w(\rho)$, is a negative integer and therefore $n \geq l + 1$ with n and l being positive integers. Accordingly, the E_n eigenenergies are in atomic units (Fig. 2.3)

$$E_n = -\frac{1}{2n^2} \quad n = \text{principal quantum number} \tag{2.50}$$

or, as mentioned before, in other units as well like when setting $\mu = m_e$)

[11]The choice is motivated by the fact that for $\rho \to \infty$, we can neglect the terms containing ρ and ρ^2 in the (2.46) obtaining the solutions $R(r) = e^{\pm\rho/2}$ (the two take only the $e^{-\rho/2}$ which vanishes at infinity). Instead, near the origin it is possible to prove that the solution must be proportional to ρ^l.

$$E_n = -\frac{m_e q^2}{2\hbar^2 n^2} = -\frac{\mu_e e^4}{2\hbar^2 n^2} \tag{2.51}$$

This gives a sequence of energy values that tend to become infinitely dense as $n \to \infty$ and $E_n \to V(r = \infty)$. To compare more favorably with experimental spectra one should replace m_e with the actual reduced mass of the system. This provides the discrete spectrum of the hydrogen atom energies. If one also includes relativistic terms and the effect of the Lamb shift [10] there is almost exact agreement between theory and experiment, which was a major triumph for quantum mechanics. In the case of l and n integers, the confluent hypergeometric function coincides, apart from a normalization factor, with the generalized (associated) Laguerre polynomials L_n^l.

Then, the wave function (denoted by the subscript n and l showing the explicit dependence on these quantum numbers) is

$$R_{nl}(r) = N_{nl}\rho^l e^{-\rho/2} L_{n+l}^{2l+1}(\rho) \tag{2.52}$$

where N_{nl} is a normalization factor, the value is

$$N_{nl} = \frac{(n-l-1)!}{[2n(n+l)!]^3}. \tag{2.53}$$

Simple recurrence relations can be used to calculate the polynomials for all values of the parameters that characterize this simple two-particle system. After all, the same formalism applies in general, with the proper tuning of the parameters (e.g., Z can be quite different from 1 and $\mu = m_1 m_2/(m_1 + m_2)$), to any ion–ion interaction to evaluate the bound states of the related diatomic molecule. Because of the conventions used, one should be careful in comparing the hydrogen atom solutions with those of a diatomic molecule. For diatomic molecules, we label the vibrational states for each l with the quantum number ν, which is not the principal quantum number used for the hydrogen atom n. The relationship between these two quantum numbers is $\nu = n - l - 1$. Then for each l the number of nodes in the wave function is equal to the vibrational quantum number and for a specified ν, l can range from 0 to ∞.

2.2.2 *The Formulation of Quantum Elastic Scattering*

A significant difference between the Coulomb and the HO potential is the fact that for the former E can be higher than $V(r = \infty)$. In this case, the wave function does not vanish at large distances and the energy values are not discrete (although they are still eigenstates) and can vary in a continuous manner from zero to infinity). Accordingly, n and ρ variables introduced in (2.45) are imaginary with important consequences on the nature of the solution of the radial equation.

In order to find the quantum solution, we must analyze the physics of the problem and define proper boundary conditions of the differential equation (2.27). For this, we

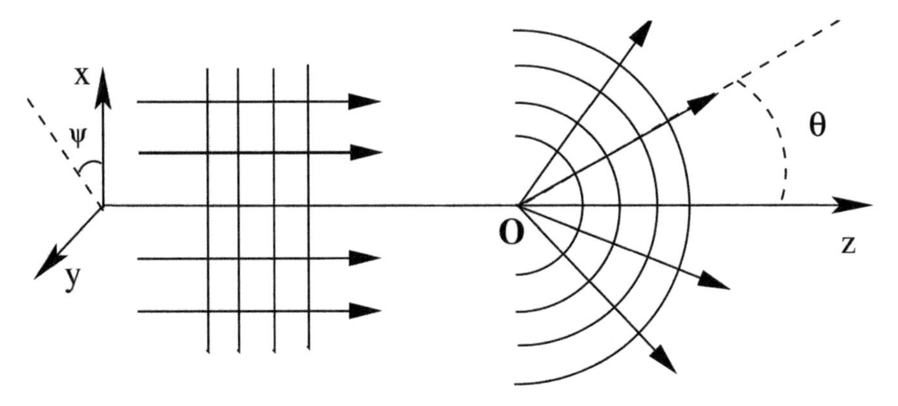

Fig. 2.4 Incident plane wave (left hand side panel) and scattered wave (right hand side panel) from the center of potential **O**

start from the case of a free particle (no potential) with momentum $\mathbf{p} = \hbar\mathbf{k}$ where $\mathbf{k}$ is the wave vector, also called the wave number, that coincides with the momentum, divided by $\hbar$, of the particle (see Eq. 2.20). The wave function that describes the free particle[12] is that of a plane wave $e^{i\mathbf{k}\cdot\mathbf{r}}$ such that

$$\tilde{\Psi}(\mathbf{r}) = e^{i\mathbf{k}\cdot\mathbf{r}} \tag{2.54}$$

In the case of a collision process the incident wave that initially describes the system has the form of a plane wave.

If you match the z-axis of the reference system (see Fig. 2.4) with the direction of $\mathbf{k}$, the incident plane wave is $\Psi_{inc}(\mathbf{r})$ $(r \rightarrow -\infty)$ is given by $e^{i\mathbf{k}\cdot\mathbf{r}} = e^{ikr\cos\theta} = e^{ikz}$ and therefore does not depend on the azimuthal angle ψ (this problem, as already seen in Chap. 1, has cylindrical symmetry).

The effect of the introduction of a potential is to disperse this plane wave transforming it asymptotically $(r \rightarrow \infty)$ into a spherical wave (scattered wave) centered at the origin (again see Fig. 2.4)

$$\Psi_{dif}(\mathbf{r}) = \frac{1}{r}e^{ikr}f(\theta) \tag{2.55}$$

where $f(\theta)$ is the scattering amplitude. It has the dimension of a length and, as we shall see below, determines the cross section of the collisional process.[13] Accordingly, the general form of the asymptotic solution will be of the form

---

[12]Or solution of the Schrödinger $H_0\tilde{\Psi} = E\tilde{\Psi}$, where H_0 is the free particle Hamiltonian of the system $H_0 = -\frac{\hbar^2}{2\mu}\nabla^2$ for positive values of energy for which $E = \hbar^2k^2/2m$ (with $k = |\mathbf{k}|$).

[13]In the case of *scattering from an anisotropic potential* the scattering amplitude does not only depend on the azimuthal angle but also on ψ.

$$\Psi(\mathbf{r}) \overset{r\to\infty}{\sim} \Psi_{inc}(\mathbf{r}) + \Psi_{dif}(\mathbf{r}) = e^{ikz} + \frac{1}{r}e^{ikr}f(\theta). \tag{2.56}$$

The boundary conditions of our problem are the following:

$$\Psi(\mathbf{r}) = 0, \quad r \to 0$$
$$\Psi(\mathbf{r}) = \Psi_{inc}(\mathbf{r}) + \Psi_{dif}(\mathbf{r}) = e^{ikz} + \frac{1}{r}e^{ikr}f(\theta), \quad r \to \infty$$

To express the amplitude $f(\theta)$ as a function of variables that can be easily calculated by integrating the equation, we rewrite Eq. 2.56 as

$$f(\theta) \simeq [\Psi(\mathbf{r}) - \Psi_{inc}(\mathbf{r})]\, re^{-ikr} \tag{2.57}$$

and, using a typical procedure of physical sciences, expand both $\Psi(\mathbf{r})$ that $\Psi_{inc}(\mathbf{r})$ in the series of the products of the radial solution $\xi(r)$ and Legendre polynomials

$$\Psi(\mathbf{r}) = \frac{1}{r}\sum_{l=0}^{\infty} A_l \xi_l(r) P_l(\cos\theta) \tag{2.58}$$

and

$$\Psi_{inc}(\mathbf{r}) = e^{ikz} = \frac{1}{r}\sum_{l=0}^{\infty} \tilde{A}_l \tilde{\xi}_l(r) P_l(\cos\theta). \tag{2.59}$$

This method is commonly referred to as partial wave expansion.

As regards to the radial function $\tilde{\xi}_l(r)$, it is the solution of Eq. (2.27) in the absence of any potential ($U(r) = 0$). To determine the form of $\xi_l(r)$, it is useful to proceed as follows. First, introduce the new variable $\rho = kr$ and then define the function $\tilde{g}_l(\rho) = \tilde{\xi}_l/\rho$ so to obtain the equation (Fig. 2.5)

$$\left[\frac{\mathrm{d}^2}{\mathrm{d}\rho^2} + \frac{2}{\rho}\frac{\mathrm{d}}{\mathrm{d}\rho} + \left(1 - \frac{l(l+1)}{\rho^2}\right)\right]\tilde{g}_l(\rho) = 0. \tag{2.60}$$

As can be seen in Ref. [9], the analytical solutions of this equation are spherical Bessel functions j_l, and Neumann functions η_l or equivalently the spherical Hankel function $h_l^{(1)}$ and $h_l^{(2)}$ (or functions Bessel, respectively, of the first, second, and third type). Consequently the relative linear combinations of these solutions, which vanishes at the origin, as required by the first of the boundary condition of the problem, is the Bessel function j_l[14] whereby

$$\tilde{g}_l(\rho) = j_l(\rho) \quad \text{or} \quad \tilde{\xi}_l(r) = krj_l(kr). \tag{2.61}$$

[14]Precisely, for $\rho \to 0$ the Bessel functions are proportional to ρ^l. Such a function is defined regular at the origin. The function of Neumann and Hankel functions are irregular at the origin. For example, the function $\eta_l(\rho) \overset{\sim}{\rho}^{-l+1}$, cannot be accepted as a solution because it is divergent at the origin.

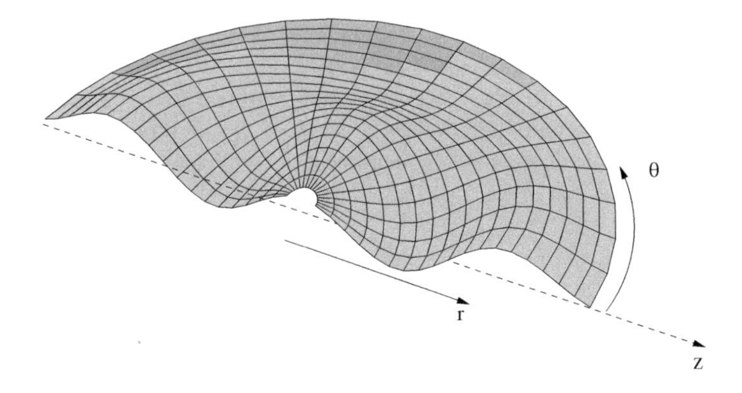

Fig. 2.5 Real part of the wave with $l = 2$: Re $\{(2l+1)i^l j_l(kr)P_l(\cos\vartheta)\}$. The graph makes use of a set of polar coordinates suitable for representing functions of the type $f = f(r, \theta)$ with $0 \le r \le \infty$ and $0 \le \theta \le \pi$

whose asymptotic form (see Ref. [9]) is

$$\tilde{\xi}_l(r) \overset{r\to\infty}{\sim} kr\frac{\sin(kr - l\pi/2)}{kr} = \sin(kr - l\pi/2). \tag{2.62}$$

It is easy, although not trivial (see Appendix A3) to determine the coefficients $\tilde{A}_l$ of the expansion (2.59). If we consider, the following expansion:

$$\Psi_{inc}(\mathbf{r}) = e^{ikz} = e^{ikr\cos\theta} = \sum_{l=0}^{\infty}(2l+1)i^l j_l(kr)P_l(\cos\theta). \tag{2.63}$$

and in the asymptotic limit

$$\begin{aligned}\Psi_{inc}(\mathbf{r}) = e^{ikz} \overset{r\to\infty}{\to} &\sum_{l=0}^{\infty}(2l+1)i^l\frac{\sin(kr - l\pi/2)}{kr}P_l(\cos\theta) \\ &= \frac{1}{2i}\sum_{l=0}^{\infty}(2l+1)i^l\left\{\frac{e^{i(kr-l\pi/2)}}{kr} - \frac{e^{-i(kr-l\pi/2)}}{kr}\right\}P_l(\cos\theta)\end{aligned} \tag{2.64}$$

comparison of the above equation with Eq. 2.59 gives $\tilde{A}_l = (2l+1)i^l/k$. This last Eq. 2.64 has a particularly important physical meaning. In fact, the plane wave and e^{ikz} can be seen as a superposition of an infinite number of spherical waves outgoing $e^{i(kr-l\pi/2)}/kr$ and incoming $e^{-i(kr-l\pi/2)}/kr$ waves. This expansion takes advantage of the fact that functions $j_l(kr)P_l(\cos\theta)$ or $j_l(kr)Y_{lm}(\theta,\psi)$ constitute a complete set.[15]

[15] A set of functions $g_1, g_2, \ldots, g_i, \ldots$ constitutes a complete set if a function f, which satisfies the same boundary conditions of the functions g_i and clearly function of the same variables, can be expressed (see [11]) as a linear combination of these functions

The superposition of states is one of the fundamental aspects upon which the entire system of theoretical quantum mechanics rests.[16]

Regarding the radial function $\xi_l(r)$ at a point that is far from the center of interaction (i.e., for significantly large values of r) if $r^2U(r) \to 0$ for $r \to \infty$ (which is not valid, for example, in the already mentioned case of the Coulomb potential), the solution will still be that of Eq. 2.61 but in its more general form

$$\xi_l(r) = kr\left[\alpha_l j_l(kr) + \beta_l \eta_l(kr)\right] \tag{2.65}$$

which is obtained from the asymptotic hypergeometric functions which take into account the effect of potential in the region where it is not negligible. The coefficients of this linear combination can be written in the following way (recalling the asymptotic expression of the Bessel and Neumann functions)

$$\begin{aligned} \xi_l(r) &= kr\left[A_l\left(\cos\delta_l\, j_l(kr) - \sin\delta_l\, \eta_l(kr)\right)\right] \\ &\overset{r\to\infty}{\sim} A_l\left[\cos\delta_l \sin(kr - l\pi/2) + \sin\delta_l \cos(kr - l\pi/2)\right] \\ &= A_l \sin(kr - l\pi/2 + \delta_l). \end{aligned} \tag{2.66}$$

The quantity δ_l is a phase shift which is the asymptotic difference between the incoming and outgoing wave. It can be positive or negative depending on whether the wave is shifted to smaller or larger radii when compared to the solution in the absence of a potential and can be computed from the asymptotic value of the radial function.

In fact, from (2.65) we have

$$-\frac{\beta_l}{\alpha_l} = \frac{\sin\delta_l}{\cos\delta_l} = \tan\delta_l. \tag{2.67}$$

So we are now able to write the asymptotic form of the full wave function:

$$\Psi(\mathbf{r}) \overset{r\to\infty}{\sim} \sum_{l=0}^{\infty} A_l \frac{\sin(kr - l\pi/2 + \delta_l)}{kr} P_l(\cos\theta) \tag{2.68}$$

(Footnote 15 continued)

$$f = \sum_i a_i g_i \quad \text{with} \quad a_i = constants.$$

According to a fundamental postulate of quantum mechanics eigenfunctions of a Hermitian operator associated with an observable physical constitute a complete set. In this sense, the functions of the HO can be used as a basis for the development of functions in many applications.

[16]For a general discussion on the aspects fundamentals of quantum mechanics, see: G.Carlo Ghirardi, *A look at the cards of God*, The Assayer, 1997 in Milan.

Equating (2.56) to (2.68) and using the asymptotic expression of the plane wave given in Eq. (2.64), canceling the incoming wave (i.e. the coefficients of the term e^{-ikr}) allows us to determine coefficients A_l (the calculation is a bit long but not difficult):

$$A_l = (2l+1)i^l e^{i\delta_l}/k. \tag{2.69}$$

With simple algebra,[17] we can then write the full asymptotic form of the wave function (see Eq. 2.68)

$$\begin{aligned}\Psi(\mathbf{r}) &\overset{r\to\infty}{\sim} \sum_{l=0}^{\infty}(2l+1)i^l e^{i\delta_l}\frac{\sin(kr - l\pi/2 + \delta_l)}{k}P_l(\cos\theta) \\ &= \frac{1}{2ikr}\sum_{l=0}^{\infty}(2l+1)i^l e^{i\delta_l}\left\{e^{i(kr-l\pi/2+\delta_l)} - e^{-i(kr-l\pi/2+\delta_l)}\right\}P_l(\cos\theta) \\ &= \frac{1}{2ikr}\sum_{l=0}^{\infty}(2l+1)\left\{\mathcal{S}_l - (-1)^l e^{-ikr}\right\}P_l(\cos\theta)\end{aligned} \tag{2.70}$$

where $\mathcal{S}_l = e^{2i\delta_l}$ is the element of the **S** matrix[18] in the case of *elastic scattering*. From Eq. 2.57, we obtain[19]:

$$\begin{aligned}f(\theta) &= \frac{2i}{k}\sum_{l=0}^{\infty}\left[A_l e^{i\delta_l} - \tilde{A}_l\right]e^{-il\pi/2}P_l(\cos\theta) \\ &= \frac{1}{2ik}\sum_{l=0}^{\infty}e^{-il\pi/2}(2l+1)i^l[\mathcal{S}_l - 1]P_l(\cos\theta) \\ &= \frac{1}{2ik}\sum_{l=0}^{\infty}(2l+1)[\mathcal{S}_l - 1]P_l(\cos\theta)\end{aligned} \tag{2.71}$$

which gives us the scattering amplitude as a function of the phase shift δ_l. Taking into account that the Legendre polynomials satisfy the relationship[20]

[17]From the formulae $\sin z = \frac{e^{iz}-e^{-iz}}{2i}$, also $e^{-l\pi/2} = (1/i)^l$, and by definition of complex number $i^{2l} \equiv (-1)^l$.

[18]The *scattering matrix* **S** plays a fundamental role (see text [5, 8]) in the scattering theory because it is the quantity linking theory and experiment.

[19]In the penultimate of the following steps, we use the trivial transformation $\frac{e^{2i\delta_l}-1}{2i} = \frac{e^{i\delta_l}(e^{i\delta_l}-e^{-i\delta_l})}{2i} = e^{i\delta_l}\sin\delta_l$.

[20]Derived from the orthonormality relationship of the Legendre polynomials, see what has been said in Appendix A3.

$$\sum_{l=0}^{\infty}(2l+1)P_l(\cos\theta) = 0 \quad \text{to } \theta \neq 0 \tag{2.72}$$

then we can rewrite the scattering amplitude as (remembering that $P_l(1) = (-1)^l$)

$$f(\theta) = \begin{cases} \frac{1}{2ik}\sum_{l=0}^{\infty}(2l+1)S_l P_l(\cos\theta) & \text{for } \theta \neq 0 \\ \frac{1}{k}\sum_{l=0}^{\infty}(2l+1)\sin\delta_l e^{i\delta_l} & \text{for } \theta = 0 \end{cases} \tag{2.73}$$

2.2.3 *The Quantum Elastic Scattering Cross Section*

Let us see how we can derive the elastic cross section from the scattering amplitude $f(\theta)$. The flux (i.e., the number of particles passing through a unit surface area in a second) is given by the wave velocity times the square wave function, i.e., $|\Psi|^2 v$ describing the collision. (remember that the square of the wave function describing the state of the system provides us with the wave intensity[21]). Since, $v = \hbar/k\mu$ (see Eq. (2.20) we have:

$$\text{incident flux} = |\Psi_{inc}(\mathbf{r})|^2 v = \hbar k/\mu \tag{2.74}$$

similarly, from the Eq. (2.56), we obtain for the outgoing flow

$$\text{outgoing flux} = |\Psi_{dif}(\mathbf{r})|^2 v = |f(\theta)/r|^2 v. \tag{2.75}$$

So the speed with which the particles spread into the unit solid $\mathrm{d}\Omega = 2\pi\sin\vartheta \mathrm{d}\vartheta$ (see Fig. 1.9) is

$$\begin{aligned} \text{speed of diffusion} &= \left|\frac{f(\theta)}{r}\right|^2 v \frac{2\pi r^2 \sin\theta \mathrm{d}\theta}{2\pi\sin\theta \mathrm{d}\theta} \\ &= |f(\theta)|^2 v. \end{aligned} \tag{2.76}$$

Then from the definition of the elastic differential cross section given in (1.53) we get:

$$\frac{\mathrm{d}\sigma}{\mathrm{d}\Omega} = \frac{|f(\theta)|^2 v}{v} = |f(\theta)|^2 . \tag{2.77}$$

which precisely provides the relationship between the cross section and the elastic scattering amplitude. Then, using (2.72) we have

[21]This statement too is one of the fundamental postulates of quantum mechanics, see [11].

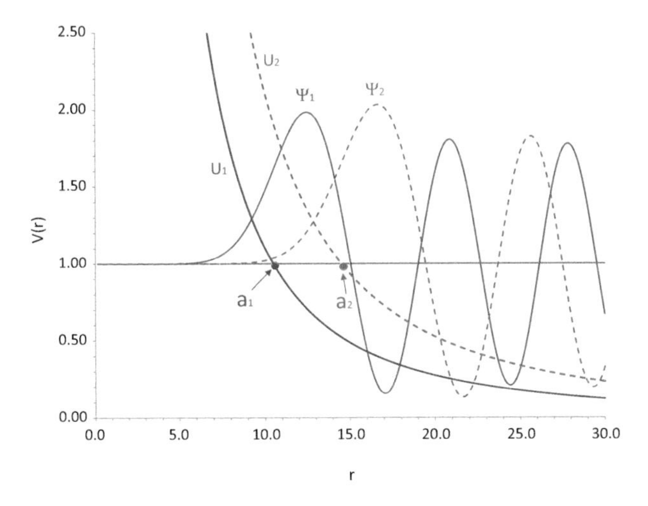

Fig. 2.6 Plot of the repulsive potentials U_1 and U_2 and of the corresponding wave functions Ψ_1 and Ψ_2 with classical turning points at a_1 and a_2

$$
\begin{aligned}
\frac{\mathrm{d}\sigma}{\mathrm{d}\Omega} &= |f(\theta)|^2 = \frac{1}{k^2}\left|\sum_{l=0}^{\infty}(2l+1)e^{i\delta_l}\sin\delta_l P_l(\cos\theta)\right|^2 \qquad (2.78)\\
&= \frac{1}{k^2}\sum_{l=0}^{\infty}\sum_{l'=0}^{\infty}(2l+1)(2l'+1)e^{i[\delta_l-\delta_{l'}]}\sin\delta_l\sin\delta_{l'}P_l(\cos\theta)P_{l'}(\cos\theta).
\end{aligned}
$$

If now we integrate over the entire solid angle, taking into account the relationship of orthogonality of the Legendre polynomials, we find that only the terms with $l = l'$ contribute to the double summation and therefore to total cross section is:

$$
\sigma_{tot} = 2\pi\int_0^{\pi}|f(\theta)|^2\sin\theta\mathrm{d}\theta = \frac{4\pi}{k^2}\sum_{l=0}^{\infty}(2l+1)\sin^2\delta_l = \frac{4\pi}{k}Imf(0) \qquad (2.79)
$$

That is each partial wave (optical theorem; see for illustrative purposes the case of a repulsive potential in Fig. 2.6) contributes to the cross section by a factor proportional to $\sin^2\delta_l$ and with a statistical weight of $(2l + 1)$.

So the maximum contribution of each partial wave

$$
\sigma_l = \frac{4\pi}{k^2}(2l+1)
$$

is obtained for values of the phase shift equal to half multiple of π ($\delta_l = (n + 1/2)\pi$, $n = 0, \pm1, \pm2, \ldots$). Conversely, waves with $\delta_l = n\pi$ do not contribute to the total cross section.

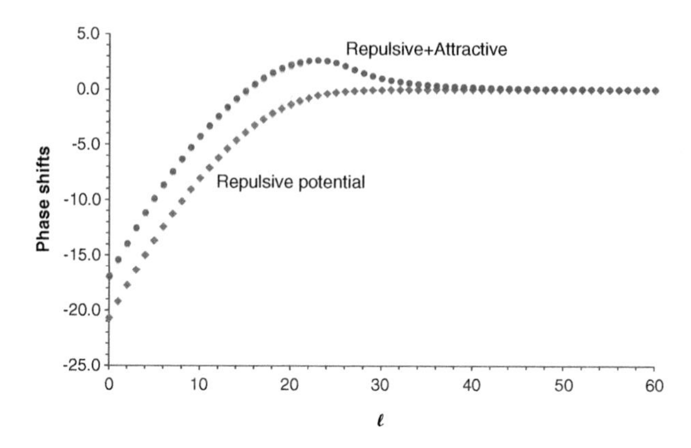

Fig. 2.7 Phase shift δ_l for a repulsive (lower curve whose wave function is plotted in Fig. 2.6 as a function of l) and an attractive–repulsive (upper curve) potential $U^l(r)$

Now, although the summation in (2.79) extends over an infinite number of partial waves, in practice only a limited number of them (though for fairly heavy systems such number may amount to several thousands) appreciably contribute to its value. In fact, for large l values the centrifugal barrier $[l(l+1)/r^2]$ (see Eq. 2.25) is such that it keeps the incident particle, or better the particle beam, out of the range of the potential $U(r)$ resulting in a negligible phase shift δ.

If we assume that there happen to be distances in excess of r_{max}, then we can estimate the maximum value of l (l_{max}) that contributes to the sum, aligning the point of return with r_{max}, i.e.,

$$\frac{l_{max}(l_{max}+1)\hbar^2}{2\mu r_{max}^2} = E$$

which leads to $l_{max} \simeq kr_{max}$.

So, to sum up, the fundamental quantity to be determined in the quantum treatment is the phase shift δ_l whose characteristics are described qualitatively for a repulsive potential in Fig. 2.7 (see the lower curve). In this case, the classical turning point a_l is greater than $\tilde{a}_l$ which involves a negative phase shift for all l.

Instead for the case of a purely attractive potential (e.g., a negative Coulomb potential) the phase shift turns out to be always positive. In the case of a potential of the attractive–repulsive type, instead, the sign of the phase shift depends on the $U(r)$ and E (like in the upper curve of Fig. 2.7) though, as l increases, the centrifugal contribution tends to dominate the contribution of the potential.

The phase shift is determined by solving the radial equation (in the following we shall discuss shortly related numerical techniques) for each partial wave l. This is the procedure commonly used at low collision energies. In the case, instead, when collision energy is large, one needs to take into account many partial waves. In this case, therefore, you may prefer to use numerical approximations (such as the JWKB) or analytical solutions (such as the Born approximation). This type of approximation falls within the type of semiclassical (SC) approximations that we also shall discuss shortly in the followings. For further details see refs. [3, 5, 12].

2.3 Realistic Models for Scattering Systems

2.3.1 *Continuum Solutions for Hydrogen-Like Atoms E > 0*

As already mentioned, with the Coulomb potential we have faced the problem of dealing for the first time with the solution for E larger than $V(r=\infty)$ that does not vanish at large distances but tends to behave as a plane wave. Accordingly, energy values are not discrete (although they are still eigenstates) and can vary continuously from zero to infinity. At the same time the variables n and ρ introduced in (2.45) are imaginary and become

$$n = -\frac{i}{\sqrt{2E}} = -\frac{i}{k} \quad \text{and} \quad \rho = 2ikr \tag{2.80}$$

where $k = \sqrt{2E}$ indicates, as usual, the wave number.

In this case the radial eigenfunctions can be formulated as

$$R_{kl} = \frac{C_k}{(2l+1)!}(2kr)^l e^{-ikr} F(i/k + l + 1, 2l + 2; 2ikr) \tag{2.81}$$

where C_K is a normalization factor and $F(i/k + l + 1, 2l + 2; 2ikr)$ is the already mentioned Hypergeometric function (see [7] for its explicit value).

However, in this case, a situation which is not uncommon in quantum treatments, having obtained a closed form solution is a bit of a Pyrrhic victory. In fact, the estimated value of the confluent hypergeometric function for arbitrary values of the arguments is not easy (and sometimes even impractical) to calculate. For this reason, even when we can give analytic closed form solutions, we often find it more convenient to determine their value using numerical techniques.

At this point, for obtaining the amplitude of diffusion and thus of total cross section it is necessary, as previously mentioned, to analyze the asymptotic behavior of the wave function (2.81). In the case of the Coulomb interaction ($rV(r) \neq 0$ as $r \to \infty$) in which the solution does not tend to Bessel functions at long range, using the asymptotic expansions of the confluent hypergeometric function (for the related rather laborious algebra we refer the reader to refs [5, 7] and exercise 111 in Ref. [13]) you get

$$\Psi(\mathbf{r}) \overset{kr\to\infty}{\sim} \sum_{l=0}^{\infty}(2l+1)i^l e^{i\delta_l} \frac{\sin(kr + \frac{1}{k}\ln 2kr - l\pi/2 + \phi_l)}{k} P_l(\cos\theta) \tag{2.82}$$

where δ_l is a phase shift formulated as

$$\delta_l = arg\Gamma(l + 1 - i/k). \tag{2.83}$$

Comparing (2.82) with its asymptotic form, we see that the only difference lies in the logarithmic term in the argument of the sin function at the numerator. In fact, in the absence of a Coulomb field,[22] $\delta_l = \arg[\Gamma(l + 1 - i/k)] = 0$. As a result Eq. (2.82) coincides with the partial wave expansion of the plane wave (see Eq. 2.64).

To the end of calculating the scattering amplitude (and therefore the cross section) we get from Eq. 2.83

$$e^{2i\delta_l} \equiv S_l = \frac{\Gamma(l + 1 + i/k)}{\Gamma(l + 1 - i/k)} \tag{2.84}$$

and obtain the resulting scattering amplitude is (2.73)

$$f(\theta) = \frac{1}{2ik}\sum_{l=0}^{\infty}(2l + 1)\mathcal{S}_l P_l(\cos\theta) \tag{2.85}$$

$$= \frac{1}{2ik}\sum_{l=0}^{\infty}(2l + 1)\frac{\Gamma(l + 1 + i/k)}{\Gamma(l + 1 - i/k)}P_l(\cos\theta) \tag{2.86}$$

By definition of the differential cross section (Eq. 1.53) in atomic units, we have

$$\frac{d\sigma}{d\Omega} = |f(\theta)|^2 = \frac{1}{4k^4}[\arcsin(\theta/2)]^4 \tag{2.87}$$

that coincides with the Rutherford formula (1.65) of the classical treatment. Let us not forget the laboratory process by which we obtained this especially interesting fact that the solution is really asymptotic only at large distances.

2.3.2 *The Rigid Sphere*

As just shown, the apparent simplicity of the bound state solution of the attractive Coulomb potential is clearly opposed to the complexity of its scattering solution. A better model to choose in order to find more intuitive connections between the shape of the interaction and the formulation of scattering quantities and to compare as well their classical and quantum solution is the rigid sphere one. The rigid sphere model is a pure scattering case (as opposed to the HO one that is exclusively a pure bound

[22]In fact, a property of the special function Γ is

$$\Gamma(l + 1 - i/k) = (l - i/k)\cdots(1 - i/k)\Gamma(1 - i/k)$$

.

state case). In fact, for $r \leq d$ the wave function has to be set equal to zero while $r > d$ the wave function coincides with that of a free particle (see Eq. 2.65)

$$R_L(r) = \alpha_l j_l(kr) + \beta_l \eta_l(kr). \tag{2.88}$$

Having the two solutions coincide at $r = d$ to cancel each other out at this point (2.88) and, using the Eq. (2.67), we obtain for the phase shift

$$\frac{j_l(kd)}{\eta_l(kd)} = -\frac{\beta_l}{\alpha_l} = \tan \delta_l \tag{2.89}$$

In the low-energy limit ($kd << 1$), it is possible to approximate[23] the previous expression as:

$$\tan \delta_l \overset{kd<<1}{\sim} \frac{(kd)^{2l+1}}{(2l+1)!!(2l-1)!!} = \frac{(kd)^{2l+1}}{(2l+1)(1\cdot 3\cdots(2l-1))^2}. \tag{2.90}$$

This (2.90) demonstrates that $|\tan \delta_l|$ decreases so rapidly as l increases that only the wave with $l = 0$ contributes significantly to the total cross section. In this case, therefore, the phase shift is simply $\tan \delta_0 \simeq \sin \delta_0 \simeq \delta_0 = -kd$. Then the calculation of the cross section according to Eq. (2.79) is immediate

$$\sigma_{tot} = \frac{4\pi}{k^2} \sum_{l=0}^{\infty} (2l+1) \sin^2 \delta_l = \frac{4\pi}{k^2} \sin^2 \delta_0 \simeq \frac{4\pi k^2 d^2}{k^2} = 4\pi d^2 \tag{2.91}$$

or four times the geometric classic value (see Eq. 1.58). This is mainly due to refraction phenomena. In fact $kd << 1$ implies $d << k^{-1} \simeq \lambda$, i.e., smaller than the De Broglie wavelength for which quantum effects are important.

It is important to note that one has

$$\alpha_l \overset{k\to 0}{\to} -\frac{\tan \delta_l}{k^l} \tag{2.92}$$

that is a constant and specifically for $l = 0$ this is called the scattering length for obvious reasons. Although $\alpha_0 \geq 0$ for the rigid sphere, for general potentials α_0 is in the range $(-\infty, \infty)$. For negative scattering lengths and in the zero-energy limit, the interaction is attractive and for positive scattering lengths the interaction is repulsive. It is very interesting to note that the scattering length can be tuned to any value with the use of an external magnetic field.

---

[23] In fact, for $z = k \to 0$ we have (see Appendix C of Ref. [5])

$$j_l(z) \overset{z\to 0}{\sim} \frac{z^l}{(2l+1)!!} \text{ and } \eta_l(z) \overset{z\to 0}{\sim} -\frac{(2l-1)!!}{z^{l+1}}$$

where $(2l \pm 1)!! = 1 \cdot 3 \cdot 5 \cdots (2l \pm 1)$.

In the high-energy limit (i.e., when $kd >> 1$) the asymptotic expressions of the Bessel and Hankel functions provide us with additional insight

$$\tan\delta_l \overset{kd>>1}{\sim} -\frac{\sin(kd - l\pi/2)}{\cos(kd - l\pi/2)} \tag{2.93}$$

where $\delta_l = -\arctan(kd - l\pi/2)$. The fact that $kd >> 1$ implies that many partial waves contribute to the total cross section. For the calculation of this, we note that the sum over l in Eq. (2.79) can be replaced by an integral[24]

$$\sum_{l=0}^{l_{max}} \simeq \int_0^{l_{max}} (2l+1)\mathrm{d}s \simeq l_{max}^2 \simeq (kd)^2. \tag{2.94}$$

In addition, the factor $\sin^2\delta_l$ in the summation that appears in the definition of the total cross section can be approximated by the average value 1/2 (approximation of the random phase), such that we will simply note[25]

$$\sigma_{tot} = \frac{4\pi}{k^2}\left(\frac{1}{2}\right)(kd)^2 = 2\pi d^2. \tag{2.95}$$

The fact that for high energies we obtain a quantum cross section which is twice the classical one is surprising because for $kd >> 1$ (or equivalently for the De Broglie wavelength $2\pi d >> \lambda$) you would expect to find correspondence with the classical limit. The origin of this discrepancy is a result of using a discontinuous potential in $r = d$ doing that the *scattering cannot be described classically*. The "extra" πd^2 factor derives from the interference between the incident and outgoing wave at small scattering angles where one cannot distinguish the incident and outgoing waves. In other words, at high energies, a contribution from diffractive wave-like nature of the quantum system adds up to the classical mechanics contribution.

2.3.3 The Morse Potential

In the previous chapter, we have discussed the validity of the Morse potential and its use for the classical treatment of atom–atom scattering. To the end of developing

---

[24]The considered approximation $l_{max} \simeq kr$ stems from the fact that the effective potential due to the rotation can be considered of the same order of the incident energy E, namely:

$$\frac{l(l+1)\hbar^2}{2mr^2} \simeq E \quad \text{or} \quad l_{max} \simeq \frac{\sqrt{2mE}}{\hbar} r \simeq kr$$

[25]The logical procedure adopted for obtaining this result is an example of a heuristic procedure.

the corresponding quantum treatment, we first write the Morse potential given there (see Eq. 1.72) as

$$U(r) = D\left\{1 - e^{-\alpha(r-r_e)}\right\}^2 - D = D(1-n)^2 - D \tag{2.96}$$

or in a reduced form as

$$U(x) = D(e^{-2\alpha x} - 2e^{-\alpha x}) = De^{-\alpha x}(e^{-\alpha x} - 2) \tag{2.97}$$

where x is the displacement from equilibrium distance $x = r - r_e$.
The related radial Schrödinger equation reads

$$\frac{\mathrm{d}^2\phi(x)}{\mathrm{d}x^2} + \frac{2\mu}{\hbar^2}\left(E - De^{2\alpha}x + 2D^{-\alpha x}\right)\phi(x) = 0 \tag{2.98}$$

In order to find the corresponding wave function, one can formulate the new variables

$$\xi = \frac{2\sqrt{2\mu D}}{\alpha\hbar}e^{-\alpha x} \quad s = \frac{\sqrt{-2\mu E}}{\alpha\hbar} \tag{2.99}$$

and adopt the following notation

$$n = \frac{\sqrt{2\mu D}}{\alpha\hbar} - \left(s + \frac{1}{2}\right). \tag{2.100}$$

Accordingly the Schrödinger equation becomes

$$\frac{\mathrm{d}^2\phi(\xi)}{\mathrm{d}\xi^2} + \frac{1}{\xi}\frac{\mathrm{d}\phi(\xi)}{\mathrm{d}\xi} + \left(-\frac{1}{4} + \frac{n+s+1/2}{\xi} - \frac{s^2}{\xi^2}\right)\phi(\xi) = 0 \tag{2.101}$$

whose solution can be formulated as

$$\phi(\xi) = e^{-\xi/2}\xi^s w(\xi).$$

By considering only the discrete spectrum of energies, $(E < V(r = \infty))$ $w(\xi)$ can be determined from the equation

$$\xi\frac{\mathrm{d}^2 w(\xi)}{\mathrm{d}\xi^2} + (2s + 1 - \xi)\frac{\mathrm{d}w(\xi)}{\mathrm{d}_{\text{,}}} + nw(\xi) = 0 \tag{2.102}$$

for which $w(\xi)$ is the confluent hypergeometric function already considered for the Coulomb potential,

$$w(\xi) = F(-n, 2s + 1; \xi).$$

From the condition that $w(\xi)$ must be finite for $\xi = 0$, and must tend to infinity for $\xi \to \infty$ no faster than a finite power of ξ, we have the equation of the spectrum of the Morse energy levels

$$E_n = -D\left[1 - \frac{\alpha\hbar}{\sqrt{2mD}}\left(n + \frac{1}{2}\right)\right]^2$$

provided that n is an integer and varies from 0 to the maximum value allowed by inequality below (please note that contrary to the Coulomb case there is always a finite number of eigenstates for the Morse potential)

$$\frac{\sqrt{2mD}}{\alpha\hbar} > n + 1/2.$$

Please note also that the solution has been obtained by extending the range of $x = r - r_e$ from $[-r_e, \infty]$ to $[-\infty, \infty]$, whose validity decreases as n increases (higher eigenstates). When $E < V(r = \infty)$, though, one can still work out the exact analytic solution for the radial Morse oscillator by taking a linear combination of Hypergeometric functions as done previously for the radial HO. This makes a negligible change in the bound state energies although it can alter the scattering wave function at high energy.

Yet, with regard to the solution of dispersion for positive energy values ($E > V(r = \infty)$) it is not possible in this case to obtain a closed-form solution for arbitrary l values (though for $l = 0$ the analytic solution is also a hypergeometric function). This leads us to considering the related numerical techniques.

2.4 Numerical Integration of the Schrödinger Equation

2.4.1 *Expectation Values of the Operators*

As already indicated, the difficulty in finding analytical solutions (in particular to the problem of dispersion) has made it necessary to develop accurate numerical methods. The numerical methods are, in fact, easily applicable to any kind of potential regardless of its complexity. Hereafter, we will focus on the methods that solve the Schrödinger equation of the following general formulation

$$\left[\hat{H}_r - E\right]\Psi(r) = 0 \tag{2.103}$$

given that this is useful to the calculation of the expectation values andor matrix elements of any generic operator $\hat{O}$. The matrix elements of a generic $\hat{O}$ operator are, in fact, defined as

$$O_{nm} = \left\langle \psi_n \left| \hat{O} \right| \psi_m \right\rangle \qquad (2.104)$$

(where ψ_m and ψ_n are the eigenfunctions of Eq. 2.103) and the expectation values of the operators in state n are the diagonal elements. The numerical approximation to a one- dimensional integral is very simple and straightforward

$$I = \int_a^b f(x)dx \approx \sum_{i=1}^{N} w_i f(x_i) \qquad (2.105)$$

where x_i are the discrete quadrature points and w_i are the related weights with N being the number of points used. We note that this sum is just a dot product between the vectors $\vec{w}$ and $\vec{f}$, that is, $I = \vec{w} \cdot \vec{f}$ which is very efficient on computers. The weights are positive definite $w_i > 0$.

There are different quadrature schemes for one-dimensional integrals. Here, we illustrate a few examples all using stepsize h:

Trapezoidal rule: The initial value of the N intervals in which the integral is partitioned is set using the $x_i = x_{i-1} + h = x_1 + (i-1)h$ relationship where h is a constant step size.[26] The weights for the extended trapezoidal rule are

$$w_i = \begin{cases} h/2 & \text{if}\, i = 1 \,\text{or}\, N \\ h & \text{otherwise} \end{cases}$$

This rule (that leads to an error twice as large if compared to the one of the standard (midpoint) trapezoidal rule of Eq. 1.50) is very easy to derive using a linear function in x to connect adjacent points.

Simpson's rule here again we are using a constant step size $x_i = x_1 + (i-1)h$ and the weights are derived from the extended trapezoidal rule as follows:

$$w_i = \begin{cases} h/3 & \text{if}\, i = 1 \,\text{or}\, N \\ 2h/3 & \text{if}\, i \,\text{is even} \\ 4/3 & \text{if}\, i \,\text{is odd} \end{cases}$$

The Simpson's rule is usually much more accurate than the trapezoidal rule since the Simpson's rule uses a quadratic function instead of a linear function to connect adjacent points. Higher order quadrature formulas may be less accurate since the interpolating function between adjacent points may ring (rapidly oscillate) especially if the function you are trying to integrate has some error associated with it (as in the case of experimental data).

[26] We note that when writing computer programs one should always use the algorithm $x_i = x_1 + (i-1)h$ (and not the recursive $x_i = x_{i-1} + h$ one) in order not to lose accuracy.

Gaussian quadratures There are many different Gaussian quadratures (i.e., Gauss–Hermite, Gauss–Laguerre, Gauss–Legendre, Gauss–Mehler to name a few). These quadratures are highly accurate since the quadrature weights and abscissas (quadrature points) are determined to make some integrals exact as follows:

$$X_{nm} \equiv \int_a^b x\psi_n(x)\psi_m(x)W(x) = \sum_{i=1}^{N} \psi_n(x_i)x_i\psi_m(x_i)w_i \tag{2.106}$$

The N abscissas x_i and N weights w_i are chosen to make these integrals exact. It should be noted that $w_i \neq W(x_i)$. The weighting function $W(x)$ is a positive definite function like e^{-x} in a Gauss–Laguerre quadrature or e^{-x^2} for a Gauss–Hermite quadrature. The $\mathbf{X}$ matrix is symmetric and hence its eigenvalues x_i are real. In fact the abscissa or quadrature points x_i are just the eigenvalues of the matrix $\mathbf{X}$. It is interesting to note that the mth eigenvector of $\mathbf{X}$ is $\sqrt{w_i}\psi_m(x_i)$ and thus one can easily evaluate the quadrature weights w_i. This method requires that we know the analytical functions $\psi_n(x)$ and that we can analytically evaluate the matrix elements in Eq. 2.106. Then one can approximately evaluate similar integrals of the form

$$I = \int_a^b f(x)W(x)dx \approx \sum_{i=1}^{N} w_i f(x_i) \tag{2.107}$$

In this case, the abscissas are not equally spaced and none of quadrature points and abscissas are the same if one goes from an N-term quadrature to a M- term quadrature. In fact, the abscissas in a N term Gauss–Hermite quadratures are the zero of $H_{N+1}(x)$. Likewise, the abscissas for a Gauss–Legendre quadrature are the zeros of Legendre polynomials, etc. These methods, therefore, do not allow the reuse of previously calculated values that is instead easy to implement when using constant step size methods by doubling the number of quadrature points.
Accordingly, we will discuss here, in more detail, only the following fixed stepsize approaches: the Laplacian operator, the wave function, and the potential.

2.4.2 *Approximation to the Laplacian*

Finite Difference

Let us first numerically approximate the Laplacian using finite differences to the differential operator. In the case of the second derivative, we have

$$\psi''(x) = \frac{\mathrm{d}^2\psi(x)}{\mathrm{d}x^2} \simeq \frac{\psi(x_{i+1}) - 2\psi(x_i) + \psi(x_{i-1})}{h^2} \tag{2.108}$$

so for any three points in the sequence one can have

$$\psi(x_{i+1}) = -\psi(x_{i-1}) + \left[2 - h^2(k^2 - V(x_i))\right] \psi(x_i) \tag{2.109}$$

as we saw previously the boundary condition at the initial point x_0 is $\psi(x_0) = 0$, which allows us to use an arbitrary value for $\psi(x_1)$, since this will only effect the normalization and not the boundary condition. Then, we propagate the solution up to the desired end point. Remember however, there are two linear- independent solutions to a second- order differential equation. One of the solutions must be regular and is the desired solution. The other solution is an irregular solution and can grow exponentially in either the positive or negative direction of propagation. This can cause serious numerical problems since the computer will initially have a very small fraction of the irregular solution. If one then propagates in the direction that the irregular solution is exponentially growing the error at each subsequent step will exponentially grow and then you will only have the irregular solution which is not the desired one. This numerical problem can be overcome by performing the propagation both from the left and from the right and then connect the two solutions at a common point making sure both solutions and their derivative are equal at that point. In the classically allowed region, the propagation is stable to propagate in either direction. However, if one propagates into the classically forbidden region one of the solutions is exponentially growing and the other is exponentially decreasing. This will determine whether you want to propagate from the left or from the right.

We can also write these equations as an eigenvalue equation: Letting

$$\mathbf{A}_{ij} = \begin{cases} -2 & \text{if } i = j \\ 1 & \text{if } i = j \pm 1 \\ 0 & \text{otherwise} \end{cases}$$

and

$$\lambda = -h^2 k^2 \tag{2.110}$$

then

$$\mathbf{Af} - \lambda \mathbf{If} = 0 \tag{2.111}$$

This is an example of a tridiagonal matrix eigenvalue problem and *LAPACK* has efficient numerical codes for solving these eigenvalue equations. Each eigenvector or column of the above eigenvalue problem is a solution to the secular equation and its associated eigenvalue. The negative eigenvalues correspond to the bound states of the system.

It is useful to understand the qualitative nature of the eigenstates of a quantum system:

- The wave function does not have any zeros in the classically forbidden regions.
- The wave function oscillates in all classically allowed regions.

- The states with energy less than the asymptotic limits of the potential are bound states. These states monotonically approach zero in the classically forbidden regions.
- The states with energy greater than the asymptotic limits of the potential are continuum states. These states always oscillate in the classically allowed regions and are characterized by a phase shift.
- The overlap between any two states with different energies and the same effective potential are orthogonal to each other:
  - All bound states are orthonormal to each other $\langle\rangle\psi_{l,E_m}|\psi_{l',E_m}\rangle = \delta_{m,n}\delta_{l,l'}$
  - All continuum states are orthogonal to each other $\langle\psi_{l,E}|\psi_{l,E'}\rangle = 0$ provided that $E \neq E'$
  - All bound and continuum states are orthogonal to each other $\langle\psi_{l,E_n}|\psi_{l',E}\rangle = 0$
- It is useful to normalize all bound states to 1
- Continuum states cannot be normalized to 1 but can be normalized in many different ways such as energy normalization $\langle\psi_{l,E}|\psi_{l,E'}\rangle = \delta(E, E')$

Numerov

Let us expand the wave function about the x_i then

$$\psi_{i+1} = \psi_i + h\psi_i' + \frac{h^2}{2!}\psi_i'' + \frac{h^3}{3!}\psi_i''' + \frac{h^4}{4!}\psi_i'''' + \frac{h^5}{5!}\psi_i''''' + \frac{h^6}{6!}\psi_i'''''' + \cdots \tag{2.112}$$

$$\psi_{i-1} = \psi_i - h\psi_i' + \frac{h^2}{2!}\psi_i'' - \frac{h^3}{3!}\psi_i''' + \frac{h^4}{4!}\psi_i'''' - \frac{h^5}{5!}\psi_i''''' + \frac{h^6}{6!}\psi_i'''''' - \cdots \tag{2.113}$$

Taking the sum of these two expressions we get

$$\psi_{i+1} + \psi_{i-1} = 2\psi_i + 2\frac{h^2}{2!}\psi_i'' + 2\frac{h^4}{4!}\psi_i'''' + 2\frac{h^6}{6!}\psi_i'''''' + 2\cdots \tag{2.114}$$

We can also do this for the second derivative giving

$$\psi_{i+1}'' = \psi''i_i + h\psi_i''' + \frac{h^2}{2!}\psi_i'''' + \frac{h^3}{3!}\psi_i''''' + \frac{h^4}{4!}\psi_i'''''' + \frac{h^5}{5!}\psi_i''''''' + \frac{h^6}{6!}\psi_i'''''''' + \cdots \tag{2.115}$$

$$\psi_{i-1}'' = \psi_i'' - h\psi_i''' + \frac{h^2}{2!}\psi_i'''' - \frac{h^3}{3!}\psi_i''''' + \frac{h^4}{4!}\psi_i'''''' - \frac{h^5}{5!}\psi_i''''''' + \frac{h^6}{6!}\psi_i'''''''' - \cdots \tag{2.116}$$

and again taking the sum of these two expressions we get

$$\psi_{i+1}'' + \psi_{i-1}'' = 2\psi_i'' + 2\frac{h^2}{2!}\psi_i'''' + 2\frac{h^4}{4!}\psi_i'''''' + 2\frac{h^6}{6!}\psi_i'''''''' + 2\cdots \tag{2.117}$$

Using (2.18) through (2.20) and solving for the second derivative of ψ we have $\psi'' = \frac{2\mu}{\hbar^2}(V^l - E) = (U^l - k^2)$ where $U^l = \frac{2\mu}{\hbar^2}V^l$ and $k^2 = \frac{2\mu}{\hbar^2}E$. Now, let us multiply the above expression by $h^2/12$ to eliminate the fourth derivative term, letting $\tau_i = \frac{h^2}{12}\left[U^l(x_i) - k^2\right]$ we obtain a two-term recurrence relation similar to what we obtained in the previous section:

$$(1 - \tau_{i+1})\psi_{i+1} - (2 - 10\tau_i)\psi_i + (1 - \tau_{i-1})\psi_{i-1} = 0 \tag{2.118}$$

This procedure is called the Numerov method and is much more accurate than the finite difference formula. It is simple to use, with a local error of h^4 instead of h^2. After taking N steps the global error is proportional to h^3, whereas the finite difference formula has a global error proportional to h.

One can also write this relation in a matrix form. Letting:

$$\mathbf{A}_{ij} = \begin{cases} -(2 + 10\tau_i) & \text{if } i = j \\ (1 - \tau_i) & \text{if } i = j \pm 1 \\ 0 & \text{otherwise} \end{cases}$$

and

$$\lambda = -\frac{h^2k^2}{12} \tag{2.119}$$

then

$$\mathbf{A}\psi - \lambda\mathbf{S}\psi S = 0 \tag{2.120}$$

which is a generalized eigenvalue problem where

$$\mathbf{S}_{ij} = \begin{cases} 10h^2/12 & \text{if } i = j \\ h^2/12 & \text{if } i = j \pm 1 \\ 0 & \text{otherwise} \end{cases}$$

and *LAPACK* has efficient numerical codes for solving the generalized eigenvalue problem.

Again, each eigenvector or column of the above eigenvalue problem is a solution to the secular equation and its associated eigenvalue. The negative eigenvalues correspond to the bound states of the system.

The Distributed Approximating Functional (DAF)

As one will shortly see, this method is considerably different from the finite difference and Numerov approximations to the Laplacian operator. We start by using the fundamental property of the Dirac delta functional $\delta(x - x')$:

$$\psi(x) = \int_{-\infty}^{\infty} \delta(x - x')\psi(x')dx'. \tag{2.121}$$

Let us make a Hermite polynomial expansion of the Dirac delta functional

$$\delta(y) = \sum_{n=0}^{\infty} a_n H_n\left(\frac{y}{\sqrt{2}\sigma}\right) e^{-\left[\frac{y^2}{2\sigma^2}\right]}. \tag{2.122}$$

where σ is a parameter determining the width of the Gaussian. Multiplying by $H_{n'}(y/(\sqrt{2}\sigma)e^{-y^2/(2\sigma^2)}$ and integrating over all space we have

$$\int_{-\infty}^{\infty} H_{n'}\left(\frac{y}{\sqrt{2}\sigma}\right) e^{-\left[\frac{y^2}{2\sigma^2}\right]}\delta(y)dy = \sum_{n=0}^{\infty} a_n \int_{-\infty}^{\infty} H_{n'}(y)H_n(y)e^{-y^2}dy \tag{2.123}$$

and then using the orthogonality of the Hermite polynominals

$$H_{n'}(0) = a_{n'}\left(\sqrt{\pi}(n')^2(n')!\right)^{-1/2}. \tag{2.124}$$

Since the odd Hermite polynomials are zero at the origin $H_{2n+1}(0) = 0$ we obtain the following expansion for the Dirac delta functional:

$$\delta(y) = \frac{1}{\sigma\sqrt{2\pi}} e^{\left[-\frac{y^2}{2\sigma^2}\right]} \sum_{n=0}^{\infty} \left[\frac{-1}{4}\right]^n \frac{1}{n!} H_{2n}\left(\frac{y^2}{\sqrt{2}\sigma}\right) \tag{2.125}$$

taking the k*th* derivative of this expression and using the recurrence relations for Hermite polynomials we have

$$\delta^{(k)}(y) = \frac{1}{\sigma\sqrt{2\pi}} \left[\frac{-1}{\sqrt{2}\sigma}\right]^k e^{\left[-\frac{y^2}{2\sigma^2}\right]} \sum_{n=0}^{\infty} \left[\frac{-1}{4}\right]^n \frac{1}{n!} H_{2n+k}\left(\frac{y^2}{\sqrt{2}\sigma}\right) \tag{2.126}$$

Now, let us define the k*th* derivative of the distributed approximating Functional approximation by terminating the infinite sum at $n = M/2$ of the Dirac delta

$$D^{(k)}_{DAF}(y) \equiv \frac{1}{\sigma\sqrt{2\pi}} \left[\frac{-1}{\sqrt{2}\sigma}\right]^k e^{\left[-\frac{y^2}{2\sigma^2}\right]} \sum_{n=0}^{M/2} \left[\frac{-1}{4}\right]^n \frac{1}{n!} H_{2n+k}\left(\frac{y^2}{\sqrt{2}\sigma}\right) \tag{2.127}$$

Then, we can also define the k*th* derivative of the wave function

$$\psi^{(k)}_{DAF}(x) = \int_0^{\infty} D^{(k)}_{DAF}(x-x')\psi_{DAF}(x')dx' \tag{2.128}$$

If we now approximate the integration via a numerical quadrature, we have

$$\psi^{(k)}_{DAF}(x) = \sum_j D^{(k)}_{DAF}(x-x_j)\psi_{DAF}(x_j)hdx' \tag{2.129}$$

which shows that the DAF can be used to find the kth derivative of the wave function at any point in space given that we know the wave function (and not its derivatives) on an equally spaced grid (the distance between the points is h in the above expression. Evaluating the expression at the same grid points we have

$$\psi_{DAF}^{(k)}(x_i) = \sum_j D_{DAF}^{(k)}(x_i - x_j)\psi_{DAF}(x_j)hdx' \tag{2.130}$$

or in matrix notation

$$\psi_{DAF}^{(k)} = \mathbf{D}_{DAF}^{(k)}\psi_{DAF} \tag{2.131}$$

where ψ is now a column vector and $\mathbf{D}$ is a square matrix. Dropping the DAF subscript for clarity gives

$$\psi^{(k)} = \mathbf{D}^{(k)}\psi. \tag{2.132}$$

Now the Schrödinger equation is an eigenvalue equation

$$(\mathbf{D}^{(2)} - \mathbf{U})\mathbf{f} - \lambda\mathbf{f} = 0 \tag{2.133}$$

where

$$\lambda = -h^2k^2. \tag{2.134}$$

Each eigenvector or column of the above eigenvalue problem is a solution to the secular equation and is associated eigenvalues. The negative eigenvalue corresponds to the bound states of the system.

Note: This DAF procedure will only work if there are significant regions where the wave function is essentially zero! It will not work for the hard sphere model potential or the Coulomb potential. For the radial wave function the DAF method only works for bound states with a sufficiently strong repulsive wall at the origin to force the wave function to be essentially zero well before $r = 0$. The DAF matrix is highly banded as a result of the Gaussian factor but is not tridiagonal like the finite difference or Numerov methods. One cannot use numerical propagation. However, there are extensions of the DAF presented here to accurately include periodic functions.

2.4.3 Approximating the Wave Function

The Finite Expansion

Let us approximate the wave function as a finite sum of N known analytical basis functions

$$\Psi(r) = \sum_{i=1}^{N} a_i f_i(r) \tag{2.135}$$

which satisfy the boundary equations. The system of N equations that results can be written as

$$\mathbf{Ha} = \lambda\mathbf{Sa} \tag{2.136}$$

where the Hamiltonian matrix elements are

$$\mathrm{H}_{ki} = \left\langle f_k(r) \middle| \hat{H} \middle| f_i(r) \right\rangle \tag{2.137}$$

and overlap matrix elements are

$$\mathrm{S}_{ki} = \langle f_k(r) | f_i(r) \rangle \tag{2.138}$$

The expansion coefficients a_i and eigenvalues λ are unknowns. One should note that the basis functions $f_i(r)$ are not necessarily orthogonal to one another. The S matrix is a metric and all of its eigenvalues must be greater than zero. Any eigenvalue of the overlap matrix S equalling zero shows that the basis functions are linearly dependent. One usually eliminates all eigenvectors whose elements are the coefficients of basis functions with eigenvalues of the overlap nearly equal to zero. This eliminates the linear dependence of the basis functions.

The equations given above is called the secular equation. Its eigenvalues are given by condition

$$det\,|\mathbf{H} - \lambda\mathbf{S}| = 0 \tag{2.139}$$

The chosen $f(r)$ functions should be sufficiently simple (yet also sufficiently similar to the true solution of the original problem) so that by truncating at N one still adequately expands all wave functions for all desired energy states. This is the variational method and leads to an eigenvalue system of N equations. It is called variational since the eigenenergies are greater than the corresponding exact energies. Also, the more basis functions one uses the closer your approximate eigenenergies will be to the exact result. The number of N equations depends on how similar the basis functions are to the exact solutions. One should think about the basis functions very carefully to minimize the computational work.

One of the most stable methods used for solving the secular equation is the Jacobi method which consists of a series of similarity transformations of the type $\mathbf{P}^{-1}\mathbf{AP}$ (recall that a vector that multiplies a matrix from left is a row vector while that multiplying a matrix from right is a column vector). Each of these transformations (also known as Jacobi rotations) is a planar rotation that eliminates one of the off-diagonal elements (whichever is greater) of the matrix $\mathbf{A}$. The *LAPACK* routines contain more efficient and more reliable alternative methods to solve the secular equation.

The Discrete Variable Approximation (DVR)

In the previous section, we see the necessity to calculate matrix elements of the kinetic energy $\mathbf{T}$ and the potential energy $\mathbf{V}$ using a set of basis functions f_i. Also, in the section on numerical quadrature we discussed the Gaussian method for doing integrals. Let us use Gaussian quadrature to calculate the matrix elements of a one-dimensional potential

$$\mathbf{V}_{ij} = \int_a^b f_i^\dagger(x)_i \mathbf{V}(x) f_j(x) dx \approx \sum_k^N f_i^\dagger(x_k) \mathbf{V}(x_k) f_j(x_k) w_k. \tag{2.140}$$

Remembering that the quadrature weights w_k are positive definite and thus the square root of these weights is also real and positive we can rewrite the above equation in a slightly different form

$$\mathbf{V}_{ij} \equiv \sum_k^N w_k^{1/2} f_i^\dagger(x_k) \mathbf{V}(x_k) f_j(x_k) w_k^{1/2}. \tag{2.141}$$

letting $\mathbf{F}(x_k) = w_k^{1/2} f(x_k)$ and writing the equation in matrix form

$$\mathbf{V} = \mathbf{F}^\dagger \mathbf{V_D} \mathbf{F} \tag{2.142}$$

where $\mathbf{V_D}$ is a diagonal matrix evaluated at the quadrature points. Provided the basis functions are orthogonal, we have

$$\delta_{ij} \equiv \sum_k^N \mathbf{F}_i^\dagger(x_k) \mathbf{F}_j(x_k). \tag{2.143}$$

Or in matrix notation

$$\mathbf{I} = \mathbf{F}^\dagger \mathbf{F} = \mathbf{F} \mathbf{F}^\dagger \tag{2.144}$$

where I is the identity matrix. In other words, the matrix $\mathbf{F}$ is unitary and therefore

$$\delta_{ij} \equiv \sum_k^N F_k^\dagger(x_i) F_k(x_j) = \sum_k^N w_i^{1/2} f_k^\dagger(x_i) f_k(x_j) w_j^{1/2}. \tag{2.145}$$

Look carefully at the preceding equation since the summation index is over the kth basis function and is not an integration. You should interpret this as a completeness relation and is the discrete version (finite basis) of the Dirac delta distribution. Now let us look at the Hamiltonian matrix elements

$$\hat{\mathbf{F}}^\dagger \mathbf{H} \hat{\mathbf{F}} = \hat{\mathbf{F}}^\dagger \mathbf{T} \hat{\mathbf{F}} + \hat{\mathbf{F}}^\dagger \mathbf{V} \hat{\mathbf{F}}. \tag{2.146}$$

Multiply this equation on the left by $\hat{F}$ and on the right by $\hat{F}^\dagger$ and then use the orthogonality relation

$$\mathbf{H} = \mathbf{T} + \mathbf{V_D}. \tag{2.147}$$

where $\mathbf{V_D}$ is a diagonal matrix and $\mathbf{T}$ is a full matrix. It is easy to see that $\mathbf{T}$ is a full matrix since $\mathbf{T}$ and $\mathbf{V_D}$ do not commute $\mathbf{TV_D} \neq \mathbf{V_D T}$ (canonically conjugate variables) and therefore $\mathbf{T}$ cannot be diagonal if $\mathbf{V_D}$ is diagonal. Then all we need to do is solve the secular equation to find the appropriate eigenfunctions $\psi_i(x_k)$ and eigenenergies E_i.

$$\mathbf{H}\psi = E\psi. \tag{2.148}$$

It should be noted that each eigenvector ψ_i is just the desired wave function evaluated at the Gaussian quadrature points instead of the coefficients of the basis functions.

In this procedure, we only need to calculate the potential at the grid points which is very beneficial especially if the potential depends on other coordinates or is difficult to calculate. It should be noted that this method is equivalent to a basis set method and how rapidly it converges depends on the basis function used.

2.4.4 The Approximation to the Potential

If we approximate the potential in each interval with a constant value then the solution is analytical and of trigonometric type. In the classically allowed regions one uses a linear combination of sin and cos whereas in the classically forbidden region one uses a linear combination of sinh and cosh. This allows us, as in the case of approximating the Laplacian, to develop a propagation method based on these analytic properties.

One can also approximate the potential as a linear function and then the appropriate solutions within an interval are airy functions $Ai(r)$ and $Bi(r)$.

In fact one can use any approximation to the potential for which there are analytic solutions.

2.5 Numerical Applications

2.5.1 Systems of Linear Algebraic Equations

The general problem of solving a system of linear algebraic equations of the type

$$\mathbf{Ax} = b \tag{2.149}$$

where $\mathbf{A}$ is the matrix of coefficients, with $\mathbf{x}$ being the vector of unknowns and $\mathbf{b}$ the vector of known terms. The following example uses the method of removal in its simplest form without any optimization

```
REPEAT irig1 FOR GOING FROM 1 TO n-1
  REPEAT FOR irig2 GOING TO BE A n +1 irig1
```

```
      VNORM = vara (irig1, irig1) / vara (irig2, irig1)
      varb (irig2) = varb (irig2) * VNORM-varb (irig1)
      REPEAT FOR ICOL GOING TO BE A n irig2
         vara (irig2, ICOL) = vara (irig2, ICOL) *
                       VNORM-vara (irig1, ICOL)
      END REPEAT ICOL
    END REPEAT irig2
END REPEAT irig1

REPEAT FOR irig1 GOING TO PASS n 1 -1
   sum = varb (irig1)
   REPEAT FOR ICOL GOING TO BE A n irig1 OF STEP -1
     sum = sum + vara (irig1, ICOL)
   END REPEAT ICOL
   varX (irig1) = sum
END REPEAT irig1
```

A popular algorithm for solving a system of linear equations is the elimination method (see Appendix A4).

2.5.2 *The Structure of the Wave Functions*

The closed-form solution of the Schrödinger equation offers us the formal instrument to define the properties of the wave functions for systems with two interacting bodies. In order to understand the structure of such functions, it is useful (and sometimes necessary) to study the zeroes of the function as well as its maximum, minimum, flex points, and so on. In the example below, we see a very simple algorithm for the numerical search of the zero points, the minima and maxima of an arbitrary function. We will assume for this purpose that the function has been well defined as a dense grid of points and stored in a vector $VALF(s)$

```
INPUT xin, xfin, n
dx=(xfin-xin)/(n-1)
x=xin-dx
ISTEP REPEAT FOR GOING FROM 1 TO n
  x=x+dx
  f(x_istep)=calculation of the value of the function x_{ISTEP}
END REPEAT ISTEP
ISTEP REPEAT FOR GOING FROM 2 TO n-1
  IF(f(x_istep)*f(x_{ISTEP+1}<0)
    PRINT 'zero range', x_istep, x_{+1} ISTEP
  IF(f(x_istep)>f(x_{1}-ISTEP.And.f(x_istep)>
    f(x_{+1}ISTEP) PRINT 'max range',ISTEP-x_{1}, x_{ISTEP+1}
```

```
    IF(f(x_istep)<f(x_{1}-ISTEP.And.f(x_istep)<f(x_{+1}ISTEP)
      PRINT 'min interval', ISTEP-x_{1}, x_{ISTEP+1}
  END REPEAT ISTEP
```

The accuracy of the location of the zero points and minima and maxima obviously depends on the number of the grid points considered within the interval under consideration. The same method can be used by halving the interval (or dividing it into smaller subintervals) in several subsequent iterations in order to refine the solutions found until you reach the desired accuracy (this method is rather slow but can be easily parallelized by subdividing the intervals).

Another well-known method is the Newton–Raphson one that employs the formula

$$x_{i+1} = x_i - \frac{f(x_i)}{f'(x_i)} \tag{2.150}$$

using the values obtained by the previous iteration for the function and its derivatives to generate the next value. Applying the same algorithm to the derivative gives the minimum and maximum points. One should be careful with the Newton–Raphson method which can often diverge or never converge.

2.5.3 *The Time-Dependent Method*

As discussed at the beginning of the chapter the integration of the time-dependent Schrödinger equation (see Eq. 2.16) is usually performed by factorizing the time-dependent part of the wave function ($\Psi(\mathbf{r}, t) = \Psi(\mathbf{r})\chi(t)$) and integrating separately the stationary Schrödinger equation (see Eq. 2.103). Alternatively, the time-dependent formalism offers the possibility of integrating Eq. 2.16 by integrating in time the equation:

$$\Psi(\{\mathbf{r}\}, t) = \hat{U}(t, t_0)\Psi(\{\mathbf{r}\}, t_0) \tag{2.151}$$

where $\hat{U}(t, t_0)$ is the time evolution operator.

This means that one can define the initial shape of the wave function (or wave packet) at time $t = t_0$ and apply the Hamiltonian operator until the wavepacket has propagated into the asymptotic region. This approach allows you to use the time variable t as a continuity variable though at the price of keeping an additional variable in the formalism. This makes the approach very simple (because it is possible to evaluate the elements of the matrix **S** (the probability is the squared modulus of the corresponding **S** element) by reiterating the application of the operator evolution $e^{-i\hat{H}\tau/\hbar}$ to the wave function of the system starting from the reagents in a predetermined initial state. The procedure is repeated until the wave function is distributed across the accessible configuration space. At each time interval, the shape of the wave function in the region of interest for the evaluation of the final arrangement is analyzed. Using the Fourier transform, the information contained in the coefficients of the expansion (time-dependent) is represented as an energy-dependent form from which you can derive the matrix **S** that can be used to evaluate any experimental information you want (distributions, cross sections and coefficients of velocity reaction).

2.6 Problems

2.6.1 Qualitative Problems

1. **Phase shifts** Carefully explain why the phase shifts should approach zero as the angular momentum parameter goes to ∞.
2. **Infinite square well** What is the $l = 0$ phase shift for the infinite square well?
3. **Negative phase shifts** Explain why the phase shifts are negative for a purely repulsive potential.

2.6.2 Quantitative Problems

1. **Positronium** Positronium is similar to the Hydrogen atom except the proton is replaced with a positron (an antielectron). The positron has the same mass of the electron but its charge is positive q_e instead of negative. When we discussed the energies of the hydrogen atom, we used the mass of the electron for the reduced mass of the system. This is a good approximation because the mass of the proton is much larger than the mass of the electron. For positronium, the masses of the two particles are identical so one must use the reduced mass. Calculate the energy levels for the first 5 values of principal quantum number $n = 1, 2, 3, 4, 5$.
2. **H_2 and HD molecules** From the vibrational spectrum (see http://webbook.nist.gov/cgi/cbook.cgi?ID=C1333740&Mask=1000), determine reasonable values of D_e and α for the Morse oscillator. Use the rotational energy spacing to determine r_e. You have completely parameterized the Morse potential for H_2. Now calculate the Morse energy levels and compare with the experimental data provided. What are the energy levels for HD?
3. **Numerov propagator** Write a computer program to solve the time-independent Schrodinger equation. To test your code set the angular momentum and the potential to zero. The exact solutions are $\sin kr$. Divide the range $r = [0, 3\pi]$ into 100 equally spaced steps. Use the initial conditions $u(0) = 0$ and $u(h) = \sin kh$ where h is the equally spaced step size and $k = \sqrt{2\mu E}$. Compare your exact solutions with the numerical solutions. Now try several different values of the angular momentum. Compare your solutions with the Riccatti–Bessel functions (Download a Bessel function routine from netlib.org). In both cases, the phase

shifts should be zero. Finally, use the potential which generated the phase shift in Fig. 2.7. How do your phases shift compare with those?

4. **Differential and integrated cross sections** Either use the calculated phase shifts from the previous problem or the phase shift provided in Fig. 2.7 to calculate the differential scattering cross section and the integrated cross section.
5. **LJ and Morse potentials** Using the LJ and Morse potentials you used in Chap. 1 (problem 3) to calculate the phase shifts, differential cross section, and integrated cross section at a scattering energy equal to $D_e/2$. Can you physically explain the differences in your results? Considering the long-range behavior (van der Waals) of the interaction between two atoms predict which potential would produce the more accurate results for small collision energies.
6. **JWKB Phase shifts** Use the supplementary FORTRAN code to calculate the JWKB phase shifts for the potentials in the previous problem. How do they compare with the Numerov results. Calculate the differential cross section and the integrated cross section.
7. **Homonuclear Diatomic molecules** When the two atoms are identical one must account for this symmetry. If the two atoms are bosons the scattering amplitude is $\left[f(\theta)+f(\theta-\pi)\right]/2$. Likewise if the two atoms are fermions the scattering amplitude is $\left[f(\theta)-f(\theta-\pi)\right]/2$. Use the Morse potential parameters of the previous two problems and then plot the differential cross sections for bosons, fermions, and nonidentical atoms.
8. **Scattering length and effective range** Calculate the s-wave ($l = 0$) scattering length and effective range of the two potentials in the previous two problems by decreasing k and extrapolating to $k = 0$.
9. **Sutherland potential** Find the analytical solution to the eigenvalues of the attractive Sutherland potential
10. **Harmonic oscillator energy levels** You need to be very careful in finding the eigenvalues and eigenfunctions of any bound state. All numerical propagators are numerically stable when the wave function is exponentially increasing or oscillating. However, when the desired eigenfunction is exponentially decreasing you encounter numerical instability. Choose the lowest energy level for a harmonic oscillator numerically propagate your solution from $x = 0$ to larger values of x. You should notice that your numerical solution follows the exact solution reasonably well until x is greater than the right turning point. Your solution will continue to follow the exact results for steps and then rapidly diverge to $\pm\infty$. You need to stop your propagation as soon as it starts to diverge. A stable way to find the eigenvalues of any bound state is to propagate from $r = 0$ to $r = r_{mid}$ and also propagate from $r = r_{max}$ to $r = r_{mid}$ and then make sure the wave function and the first derivative of the wave function are equal at $r = r_{mid}$. In this way, the propagators are always propagating in the numerically stable direction.
11. **Morse oscillator energy levels** Follow the hints given in the previous problem to determine the eigenvalues and eigenfunctions of the Morse oscillator.
12. **Distributed Approximating Function—DAF** The supplementary material has FORTRAN subroutines to calculate the DAF approximate for the kinetic energy.

The potential energy term is a diagonal matrix where the *ith* diagonal is just $V(r_i)$. Use the Morse potential of the previous two problems and a matrix diagonalization routine from *www.netlib.org* to calculate all bound eigenvalues of this potential. Compare with the exact results. Plot the elements of the eigenvectors. Can you explain what these eigenvectors represent?

13. **Discrete Variable Representation—DVR** The supplementary material has FORTRAN subroutines to calculate the DVR approximate for the kinetic energy. The potential energy term is a diagonal matrix where the *ith* diagonal is just $V(r_i)$ (The supplied material also determines the nonuniformly spaced values of r_i). Use the Morse potential of the previous two problems and a matrix diagonalization routine from *www.netlib.org* to calculate all bound eigenvalues of this potential. Compare with the exact results and the DAF results if you did the previous problem.

Chapter 3
Ab initio Electronic Structure for Few-Body Systems

This chapter focuses on the problem of calculating the electronic structure of few-body systems by concentrating mainly on methods based on independent particles wavefunctions. Accordingly, we describe the variational principle, the Hartree–Fock (HF) and self-consistent field (SCF) molecular orbital models. An overview of the use of post HF configuration interaction (CI), multiconfiguration (MC) SCF, perturbation methods and density functional theory (DFT) is also given. Finally, the most popular techniques used for fitting full range potential energy surfaces suited to support the study of chemical and more specifically reactive processes are also examined.

3.1 Structured Bodies

3.1.1 The One-Electron Wavefunction Approach

In the previous chapters, we focused our attention on the case of two interacting particles using both a classical (Chap. 1) and a quantum (Chap. 2) treatment by making always the explicit assumption of their structureless nature. On this ground, we worked out the ab initio bound quantum solutions of the Schrödinger equation for the electron as a single-particle system subject to a Coulomb potential.

However, in order to generalize the ab initio treatment to molecular processes (and, in particular, to chemical reactions in which several electrons intervene and get redistributed around the related nuclei) some simplifying assumptions made in Chap. 2 (such as that collision partners are spinless, do not fragment, are subject to a central potential, etc.) have to be removed. As a consequence, the Hamiltonian of the related Schrödinger equation has to be modified accordingly to take into account that we deal with the problem of one or more electrons considered as independent (or one-electron) particles. It is worth mentioning here that even when dealing with

A. Laganà and G. A. Parker (eds.), *Chemical Reactions*, Theoretical Chemistry and Computational Modelling, https://doi.org/10.1007/978-3-319-62356-6_3

more than one nucleus we shall omit the nucleus-nucleus interaction $\hat{V}_{nn}$ term that is included in the nuclear part.

Accordingly, here we deal only with the terms associated with the interaction of the electrons among themselves and with the nuclei. This means that the additional complexity of the theoretical treatment depends not only on the inclusion of more particles (with the consequent inapplicability of the pure central field) but also on the fermionic nature of the electrons that induces a change of sign of the wavefunction following the change of assignment of the electrons to the related wavefunctions. As a matter of fact, in the remainder of the section, we shall discuss the details of the theoretical treatment of multi-electron and the multi-atom features of the potential energy surfaces governing reactive chemical processes and with their implications on the motion of the nuclei.

It is, in fact, important to stress out here that while the detail of the ab initio techniques provided in the book is instrumental to the understanding of the utilization of their outcomes in dynamical studies it would have been impossible to confine the discussion of the subject to a mere provision of some key literature references and to the listing of the most popular computer programs (see for example [1, 14–17]). Accordingly, the discussion given in this book of the most popular ab initio techniques will be directed toward the motivation of either the direct utilization of their outcomes for dynamical calculations or the construction of appropriate functional forms for fitting high level of theory potential energy values.

For this reason, we anticipate from the next chapter that here we deal only with the pure electronic wavefunction component $\Phi(\mathbf{r})$ of the total wavefunction Ψ of the molecular (electrons + nuclei) system described there due to the adoption of the Born–Oppenheimer approximation (that separates the nuclear subproblem from the electronic one). We also mention that here we use $\mathbf{r}$ as the electronic coordinate (of dimension 3 for a single electron and $3K$ for K electrons).

3.1.2 Quantum Monte Carlo

The first method to mention, thanks also to the fact that most reactions occur on the fundamental potential energy surface, is the quantum Monte Carlo (QMC) [18, 19] one. QMC calculations can produce reliable numerical solutions of the quantum many-body problem by the direct numerical integration of the many-electron multidimensional Schrödinger equation based on repeated random sampling. Its Metropolis–Hastings algorithm is also a method for obtaining a sequence of random samples from a probability distribution for which direct sampling is difficult.[1]

[1]The Metropolis–Hastings algorithm works by generating a sequence of sample values in such a way that, as more and more sample values are produced, the distribution of values more closely approximates the desired distribution, $P(x)$. These sample values are produced iteratively, with the distribution of the next sample being dependent only on the current one. Specifically, at each iteration, the algorithm picks a candidate for the next sample value based on the current one. Then, with some probability, the candidate is either accepted (in which case the candidate value is used in

The different versions of the quantum Monte Carlo method all share the common use of the technique to build accurate estimates to the multidimensional integrals that arise in the different formulations of the many-body problem. The quantum Monte Carlo methods allow for a direct treatment and description of complex many-body effects encoded in the wavefunction and offer numerically accurate solutions of the many-body problem. In principle, any physical system can be described by the many-body Schrödinger equation provided that the constituent particles are not moving at a speed comparable to that of light and that, therefore, relativistic effects can be neglected.

There are, indeed, two versions of the Monte Carlo technique that have been applied to electronic structure problems: the variational (VMC) and the diffusion (DMC) one.

In the QMC method we start from the consideration that the expectation value of the many-electron (say K) wavefunction of the ground state reads (for the equivalence between function and vector notations see Appendix A1):

$$E_0 = \frac{\langle \Phi_0 | \mathcal{H} | \Phi_0 \rangle}{\langle \Phi_0 | \Phi_0 \rangle} = \frac{\int \Phi_0^*(\mathbf{r}) \mathcal{H} \Phi_0(\mathbf{r}) \mathrm{d}\mathbf{r}}{\int \Phi_0^*(\mathbf{r}) \Phi_0(\mathbf{r}) \mathrm{d}\mathbf{r}} \tag{3.1}$$

in which $\mathbf{r}$ is the $3K$ dimensional vector of electronic positions.

The energy associated with the tentative function Φ_T is

$$E_T = \frac{\int \Phi_T^*(\mathbf{r}) \mathcal{H} \Phi_T(\mathbf{r}) \mathrm{d}\mathbf{r}}{\int \Phi_T^*(\mathbf{r}) \Phi_T(\mathbf{r}) \mathrm{d}\mathbf{r}}.$$

According to the variational principle, E_T is an upper limit to the (true) ground state energy E_0. By reformulating the integral as follows:

$$E_T = \frac{\int |\Phi_T(\mathbf{r})|^2 \frac{\mathcal{H}\Phi_T(\mathbf{r})}{\Phi_T(\mathbf{r})} \mathrm{d}\mathbf{r}}{\int |\Phi_T(\mathbf{r})|^2 \mathrm{d}\mathbf{r}}. \tag{3.2}$$

the VMC Monte Carlo method is then applied using the Metropolis–Hastings algorithm and a set of $\mathbf{r}$ values are generated in configuration space and at each of these points the energy (where $\mathcal{H}\Phi_T(\mathbf{r})/\Phi_T(\mathbf{r})$ is the "local energy") is generated. For a sufficiently large sample of points, the average value is given by

$$E_{VMC} = \frac{1}{K} \sum_{i=1}^{K} \frac{\mathcal{H}\Phi_T(\mathbf{r}_i)}{\Phi_T(\mathbf{r}_i)}. \tag{3.3}$$

---

the next iteration) or rejected (in which case the candidate value is discarded, and current value is reused in the next iteration). The probability of acceptance is determined by comparing the values of the function $f(x)$ of the current and candidate sample values with respect to the desired distribution $P(x)$.

In the VMC method, therefore, it is crucial the choice of the tentative $\Phi_T(\mathbf{r})$ that determines the value of the observable computed by the simulation.

On the contrary, the DMC approach avoids the need for setting an approximate tentative wavefunction by resorting to the use of the time-dependent formalism

$$\mathcal{H}\Phi(\mathbf{r},\tau) = -\frac{\partial\Phi(\mathbf{r},\tau)}{\partial\tau} \tag{3.4}$$

where τ is the imaginary time ($\tau = it$). In Eq. 3.4 $\Phi(\mathbf{r},\tau)$ tends to relax to the true ground state wavefunction as $\tau \to \infty$ even when one begins with any arbitrary function at $\tau = 0$ ($\Phi(\mathbf{R}, \tau = 0)$). In other words, the DMC trial wavefunction converges in any case to the numerically accurate solution and its propagation is only a mathematical expedient when the algorithm is accurate and efficient. Accordingly, one can drop τ from the notation when not strictly necessary. In addition, the method is of particular interest for distributed computing because the computational scheme can be arranged in a way that each computing node carries out an independent task (a copy of the Cambridge Quantum Monte Carlo Casino code is available at http://vallico.net/casinoqmc/).

3.1.3 *Many-Electron Wavefunctions*

Yet, the most popular approach to the problem of producing many-electron wavefunctions, starts from the corresponding single electron (hydrogen-like or their variants) solution of the time-independent Schrödinger equation given in Eq. 2.18. In fact, even if the single electron hydrogen-like wavefunctions computed in this way suffer severe limitations, they provide us with a conceptual road map that has actually led us in the past to construct many-electron wavefunctions of increasing accuracy. This is obtained by progressively integrating the Hamiltonian with the terms necessary to describe the additional complexity of molecular processes including reactions. For example, it has to be noted here that the two-body (nucleus + electron) system considered in Eq. 2.18 is inadequate to describe many-electron systems due to the insufficiency of the conservation of the classical angular momentum alone for quantum systems. This means that an additional component **s** (and the related *spin* quantum number valued either 1/2 or $-1/2$) needs to be introduced in the definition of the electronic wavefunction due to the fermionic nature of electrons. Accordingly, we shall consider the χ_{nlms} wavefunctions (the spinorbitals) associated with electrons assigned to "stable orbits of discrete energies" and characterized by the set of four quantum numbers (n, l, m and s), with no two of them having the same set of four values, as the basic building blocks of any quantum formulations of the chemical processes.

This already allows us not only to correctly describe the manifold of the excited atomic electronic states but also provides us with a proper formulation of the many-electron (say K-electronic) wavefunctions $\Phi(\mathbf{r})$ as a product of the χ_{nlms} spinorbitals.[2] In fact, to the end of extending the Hamiltonian $\hat{H}$ of Eq. 2.18 to the case of several electrons and nuclei we have to sum the terms related to the interaction of each electron with the N nuclei and the other $K-1$ electrons which read:

$\hat{T}_e = -\sum_i \frac{1}{2}\nabla_i^2$ (electronic kinetic),
$\hat{V}_{ne} = -\sum_{i,a} \frac{z_a}{r_{ai}}$ (nuclear-electronic potential)
$\hat{V}_{ee} = \sum_{i>j} \frac{1}{r_{ij}}$ (electronic–electronic potential)

for a molecule at a frozen geometry (whose coordinates, as already mentioned, are omitted for simplicity when not strictly necessary).[3]

If we enforce the drastic approximation that $\hat{V}_{ee} = 0$ for each electron, it still remain the single electron term of the sum in $\hat{T}_e$ and $\hat{V}_{ne}$. Therefore, by separating the (uncoupled) variables we have for each electron at negative energy values (i.e., lower than the corresponding asymptote) the corresponding one-electron discrete eigenfunctions χ_{nlms} that describes the probability amplitude for the electron in a hydrogen-like atom and the associated energies are expressed by the related simple analytic formulae. These mono-electronic three-dimensional wavefunctions are as usual formulated in terms of spherical polar coordinates i.e., a radius r (the distance of the electron from the nucleus) and two related angles (ϑ and ψ) for both the ground and the excited states. In case one has to assign more than one-electron, the fact that some orbitals might be occupied has to be taken into account. For example, in the case of a two-electron system, one can assign the first electron to the first spinorbital (say the lowest in energy according to the Aufbau rule) and the second electron the next in energy spinorbital. Yet, we could have chosen to assign the electrons in a different order (for example, the second electron to the first spinorbital and the first electron to the second one). This procedure has, however, to be carried out in compliance with the Pauli antisymmetry principle which requires the change of sign of the function when any two electrons are exchanged (with p being the number of permutations performed). In the simple case of the He atom (two electrons) in the

---

[2]When $\hat{H} = \hat{H}_i + \hat{H}_j$ and one has $\hat{H}_i\chi_i = E_i\chi_i$ and $\hat{H}_j\chi_j = E_j\chi_j$ the solution of $\hat{H}\chi$ can be either $\chi_i\chi_j$ or $\chi_j\chi_i$ or $\chi_i\chi_j \pm \chi_j\chi_i$.

[3]Actually the general formulation of the three terms for a system of K electrons and N nuclei should read

$$-\sum_{i=1}^{K} \frac{h^2}{8\pi^2 m_e}\nabla_i^2 \tag{3.5}$$

$$-\sum_{i=1}^{K}\sum_{a=1}^{N} \frac{Z_a e^2}{4\pi\epsilon_0 r_{ai}} \tag{3.6}$$

$$\sum_{i=1}^{K-1}\sum_{i=j+1}^{K} \frac{e^2}{4\pi\epsilon_0 r_{ij}} \tag{3.7}$$

with m_e being the electron mass. Most often, however, the use of atomic units bohr ($a_0 = \frac{h^2\epsilon_0}{\pi m_e e^2}$) for distances and hartree ($E_h = \frac{e^2}{4\pi\epsilon_0 a_0}$) for energies is found to be more convenient. Energies are accordingly expressed in hartree valued about 27.21 eV and 2626 kJ/mol.

closed shell ground state, one can obtain, therefore, both the $\chi_{1s\alpha}(r_1)\chi_{1s\beta}(r_2)$ and the $-\chi_{1s\alpha}(r_2)\chi_{1s\beta}(r_1)$ products as well as their sum.[4]

The ground state He antisymmetric wavefunction takes, therefore, the determinantal form

$$\Phi_{He} = \frac{1}{\sqrt{2!}} \begin{vmatrix} \chi_{1s\alpha}(\mathbf{r}_1) & \chi_{1s\beta}(\mathbf{r}_1) \\ \chi_{1s\alpha}(\mathbf{r}_2) & \chi_{1s\beta}(\mathbf{r}_2) \end{vmatrix}$$

or more compactly

$$\Phi_{He} = |\chi_{1s\alpha}(\mathbf{r}_1)\chi_{1s\beta}(\mathbf{r}_2)| = \frac{1}{\sqrt{2}} \left[\chi_{1s\alpha}(\mathbf{r}_1)\chi_{1s\beta}(\mathbf{r}_2) - \chi_{1s\alpha}(\mathbf{r}_2)\chi_{1s\beta}(\mathbf{r}_1)\right] \quad (3.10)$$

that for a generic K electron atom can be written as

$$\Phi = \frac{1}{\sqrt{K!}} \sum_P (-1)^p \hat{P}(\Pi_i(\phi(\mathbf{r}_i))) \quad (3.11)$$

once the possible $K!$ permutations are taken into account and the function is normalized.

For the individual spinorbital χ the spin component s (as already mentioned s can assume only the $\pm 1/2$ value) is multiplied by the analytic hydrogen-like function whose parameters ζ in the exponential part are optimized to describe the observable properties of the atom considered. More in detail, the functional form usually taken for the radial and angular component of such functions is the so-called STO (Slater Type Orbital) defined as $\chi_{STO} = Cr^{n-1}e^{-\zeta r}Y_{lm}$ due to Slater with n, l, m being the usual quantum numbers of the hydrogen-like eigenfunction. In the STO C is the normalization coefficient and Y_{lm} is the spherical harmonic in the angles ϑ and ψ. Other functional forms have become very popular for replacing the STOs because their integrals are faster to evaluate (the number of possible spinorbitals is as high as 362,880 already in the F case). The STO-NGs, in fact, have a Gaussian form in which the radial exponential is replaced by a Gaussian (e.g., $\chi_{GTO} = Ce^{-\alpha r^2}Y_{lm}$ or

---

[4]When considering a system of K electrons, if each of them is described by means of an individual one-particle function and the total system is described by means of a normalized (a wave function Φ is normalized by imposing that $\int \Phi^*\Phi d\tau = 1$ with Φ^* being its complex conjugate) product of all one-particle functions of the type

$$|\Phi\rangle = \frac{1}{\sqrt{K!}} \left[\chi_1(r_1)\chi_2(r_2).....\chi_i(r_i)\chi_j(r_j).....\chi_k(r_k)\right] \quad (3.8)$$

we can exchange the coordinates of two particles i and j as follows:

$$\hat{P}_{ij}|\Phi\rangle = \frac{1}{\sqrt{K!}}(-1)^p \left[\chi_1(r_1)\chi_2(r_2).....\chi_i(r_j)\chi_j(r_i).....\chi_k(r_k)\right] \quad (3.9)$$

with $\hat{P}_{ij}$ being a permutation operator that in the case of the electronic system preserves all of the physical properties (electrons are indistinguishable) but the sign for odd values of the parity p.

$\chi_{GTO}^{nl} = Cr^{2n-l-2}e^{-\alpha r^2}Y_{lm})$ and l, and m are not the above mentioned usual quantum numbers. As apparent from the name in order for the STO-NG to maintain the same accuracy as the STO a larger number (N) of Gaussian functions are used (STO-3G and STO-4G). In that case, one optimizes αs to fit the exponential part of the STOs.

3.1.4 The Electronic Structure of Molecules

When considering the electronic structure of molecules[5] the only practical approach is the adoption of a numerical procedure that, starting from a limited set of tentative independent particles wavefunction, generates a succession of solutions possibly converging to a sufficiently accurate many-electron wavefunction. The method of election for working such a succession is to use the variation method that minimizes the electronic energy E_e using the variational principle

$$\frac{\langle \Phi^*(\mathbf{r})|\hat{H}|\Phi(\mathbf{r})\rangle}{\langle \Phi^*(\mathbf{r})|\Phi(\mathbf{r})\rangle} = E_e \geq E_0 \tag{3.12}$$

with E_0 being the exact (unknown) energy value and $\Phi(\mathbf{r})$ being a proper antisymmetric $3K$-dimensional wavefunction. In Eq. 3.12 $\Phi(\mathbf{r})$ can be expanded in any resonable approximation to the independent particles wavefunction $\phi(\mathbf{r})$ more accurate than the above spinorbitals χ_{nlms} generated by ignoring the electron–electron interaction.

Let us, for example, assume that the wavefunctions ϕ are real, calculated using the one-electron Hamiltonian $\hat{H}$ of Eq. 2.18 including the electron–electron interaction $\hat{V}_{ee}$, normalized and mutually orthogonal. This implies that the following relationships hold

$$S_{ij} = \int \phi_i(\mathbf{r}_1)\phi_j(\mathbf{r}_2)d\tau = \delta_{ij} \tag{3.13}$$

$$E_e = \sum_{i=1}^{K} h_{ii} + \frac{1}{2}\sum_{i=1}^{K}(J_{ii} - K_{ii}). \tag{3.14}$$

with

$$h_{ii} = \int \phi_i(\mathbf{r}_1)\left(-\frac{1}{2}\nabla_i^2 - \sum_{a=1}^{N}\frac{Z_a}{r_{ai}}\right)\phi_i(\mathbf{r}_1)d\tau_1 \tag{3.15}$$

being the contribution of one-electron terms to energy (where electron 1 has been arbitrarily assigned to orbital ϕ_i and $d\tau_1$ represents the integration with respect to

[5]Although we only considered the equilibrium geometry in the above formulation one can also use the same equations at any fixed nuclear geometry. In that case it is more convenient to include in the notation the fixed nuclear geometry R as a fixed parameter in the notation.

the coordinates of electron 1),

$$J_{ii} = \sum_{j=1}^{K} \int \phi_i(\mathbf{r}_1)\phi_j(\mathbf{r}_2)\frac{1}{r_{12}}\phi_i(\mathbf{r}_1)\phi_j(\mathbf{r}_2)d\tau_{12} \tag{3.16}$$

being the Coulomb integrals (where $d\tau_{12}$ represents the integration with respect to the coordinates of electrons 1 and 2) and

$$K_{ii} = \sum_{j=1}^{K} \int \phi_i(\mathbf{r}_1)\phi_j(\mathbf{r}_2)\frac{1}{r_{12}}\phi_i(\mathbf{r}_2)\phi_j(\mathbf{r}_1)d\tau_{12} \tag{3.17}$$

being the exchange integrals.

The requirement that electronic energy E_e is a minimum according to the variational principle ($\delta E_0 \equiv \langle \Phi_0|\mathcal{H}|\Phi_0\rangle = 0$) and to the mutual orthogonality of atomic orbitals ($\langle \phi_i|\phi_j\rangle$)[6] leads to the Hartree–Fock (HF) equations

$$\hat{F}_i(\mathbf{r}_1)\phi_i(\mathbf{r}_1) = \epsilon_i\phi_i(\mathbf{r}_1) \tag{3.18}$$

in which $\hat{F}_i$ the Fock operator that can be built out of the quantities given in Eqs. 3.15, 3.16, and 3.17. Equation 3.18 provides us with the iterative mechanism in which, starting from an educated guess set of initial trial orbitals, we can at each step generate a new set of orbitals from which build a new Fock operator. The energy and/or orbital convergence values are known as self consistent field (SCF) ones. This SCF technique has been in the years analyzed and improved with respect to the nature of the orbitals, of the optimized energy values, of the convergency criteria, of the relationships with some physical observables, etc.[7]

As we shall discuss later, the HF formalism illustrated here will be taken as the ground for discussing the PES fitting techniques of relevance for reactive studies. However, in the next section we shall mention some post HF developments on which

---

[6] The minimization of a constrained function is usually carried out by the method of Lagrange multipliers (for a formal derivation see [20]).

[7] A molecule (as well as a many-electron atom) in a defined state of energy is in an eigenstate of the hamiltonian $\hat{H}$ that commutates with the angular momentum operator. This does not apply to the potential energy operator $\hat{V}$ unless it is spherically symmetric. This means that the assumption of a well-defined s, p, d, and f nature of electrons (appropriate in isolated atoms) it is not so for molecules. In this respect two lines of modeling have been developed both starting from hydrogen-like atomic functions:

(a) the molecular orbitals (MO) one combining atomic orbitals into new functions by linear combinations
(b) the valence bond (VB) one retaining the original shape of the atomic orbitals and focusing on regions of overlap to construct chemical bonds resorting, when is the case, to promotion of electrons to states of similar energy and hybridizing the involved states into an equal number of equivalent ones.

further higher levels of theory have been built. The outcomes of these higher level of theory calculations will instead be taken as electronic structure values to fit.

3.2 Higher Level Ab initio Methods

3.2.1 Beyond the Hartree–Fock Method

As already mentioned, merits and demerits of HF calculations have been repeatedly analyzed in the literature to the end of devising corrective actions. In fact, while the HF total energy represents more than 90% of the nonrelativistic one the differences in energy coming into play in chemical reactions may be of the order of 100 or 1000 times smaller. As a matter of fact when a chemical process implies the breaking/forming or even the simple exploration of potential energy regions ranging from long to short distances (with the consequent strong perturbation of the electronic distribution) one has to incorporate in the ab initio calculation the electronic correlation.

Let us start with a list of peculiarities of the HF method:

- the application of the variational principle and of the orthogonality one meets the basic accuracy criteria only for the ground state
- the SCF nature of the HF method guarantees that the electrostatic potential described by the solution wavefunctions is the same as that of the operator
- the wavefunctions not appearing in the final composition of the operator are virtual
- the energy of the HF determinant differs from that of the orbitals; their sum is the energy corrected to the next order
- Open and closed shell HF solutions are different and require two different ways (unrestricted and restricted) of dealing with the total spin.

A first point to make is that the original Fock equations were designed for atomic systems (that is for systems of spherical symmetry). In 1951 Hall and Roothaan [21] formulated the molecular spinorbitals ϕ_i as a linear combination of m basis functions

$$\phi_i = \sum_{q=1}^{m} c_{iq} \chi_q \tag{3.19}$$

transforming so far the eigenvalue problem into that of a set of algebraic equations and its solution into that of a matrix diagonalization. The selection of appropriate initial χ_q functions is, therefore, crucial for the SCF calculation. A standard approach is that of the LCAO method in which the basis set is chosen, as already mentioned, to be a linear combination of atomic orbitals centred on the various atoms. The minimum basis set to adopt would include, accordingly, the functions associated with valence orbitals.[8]

As we have already seen, the reformulation of STO atomic orbitals in terms of Cartesian Gaussian ones STO-NG is used to increase the computational efficiency of the calculations. However, in order to achieve reasonable accuracy one would have to take into account of two times their number. This type of basis is called DZ (*Double Zeta*).[9] Further improvement can be obtained, for example, by adopting the basis DZP (*Double Zeta Polarization*) that includes orbitals having higher values of l and thus takes into account the orbital distorsion due to polarization. However, in order to approach the HF limit it is necessary to use an even larger basis set. For example, in order to deal with highly excited states, diffuse Rydberg functions need to be added.

3.2.2 *The CI and MC-SCF Methods*

The method of election for recovering the residual (correlation) energy missing in the HF treatment is, however, to be found outside the blind extension of the number of HF orbitals and rely on descriptors of the instantaneous interactions between electron pairs. In other words, the main limitation of the HF theory is the neglecting of the correlation among electronic motion even if the SCF method accounts for interelectronic repulsion through the already defined Coulomb and exchange terms. Obviously, a single determinantal function (that is a single HF configuration) cannot represent the true eigenfunction even for closed-shell molecules. This type of correlation (coulomb interaction) is known as dynamic correlation and the difference between the HF energy and the unknown, E_{exact}, true one

$$\mathrm{E}_{exact} = \mathrm{E}_{HF} + \mathrm{E}_{correlation}. \tag{3.20}$$

is, indeed, the correlation energy. Two methods have been proposed in order to cope with that problem configuration interaction (CI) and multiconfiguration self consistent field (MCSCF).

In the CI method, the molecular wavefunction is a multiconfigurational (multireference) function that is expanded as a sum of Slater determinants

$$\Phi = \sum_I C_I \Phi_I \tag{3.21}$$

in which the C_I coefficients are determined using a variational procedure (i.e., by minimizing the total energy).

[8]As an example for CH_4 the minimal set is given by the functions 1s, 2s e 2p of C and 1 s of H.

[9]e.g., For the CH_4 molecule one would take, accordingly, two atomic orbital of each type.

In a full CI approach all of the excited states contributing in the sum (3.21), that is extended to all the configurations, can be built by taking into account all the possible electronic excitations to the different molecular orbitals of the system. A full CI wavefunction will consist of a large number of SCF Φ_I terms. For example, for the H_2O molecule, using a DZ basis the full CI eigenfunction will consist of about 250,000 configurations.

A full CI calculation is possible only for fairly small systems. For larger systems, one truncates the list of considered configurations to those contributing significantly to the wavefunction.

A sufficiently simple CI scheme is the CISD (*single and double configuration interaction*) one in which only the singly and doubly excited configurations are included. However, a truncated CI is not "*size consistent*".[10]

In the alternative MC-SCF (*multiconfiguration*-SCF) method a simultaneous optimization of both the C_I coefficients of (3.21) among the various configurations and the c_{iq} coefficients of the (3.19) (for example, of the LCAO expansion) is made. The simultaneous optimization of both sets of coefficients makes the MC-SCF procedure extremely heavy. For this reason, quite often in the (3.21) expansion, one considers only the configurations obtainable from a limited optimized number of molecular orbitals (active or valence orbitals). The CASSCF (*Complete Active Space*) method minimizes the number of configurations to be considered by dividing the molecular orbitals in three sets: the two inactive sets (the extreme cases of either the doubly occupied orbitals or the nonoccupied ones in all the configurations) and the active set of the intermediate orbitals occupied only at certain configurations (for small molecules the valence orbitals or the antibonding ones).

3.2.3 Perturbation Methods

An alternative method is the perturbative [22] one in which an explicit construction of the Hamiltonian matrix is needed. In this approach, the Hamiltonian operator is written as a sum of two terms

$$H = H^{(0)} + \lambda V \tag{3.22}$$

where $H^{(0)}$ is the unperturbed (zeroth order) Hamiltonian while V is a perturbation term modulated by the λ coefficient. Then for the i-th state the wavefunction Φ_i and the energy E_i are expanded in terms of subsequent corrections $\Phi_i^{(n)}$ and $E_i^{(n)}$ as follows:

$$\Phi_i = \sum_{n=0} \lambda^{(n)} \Phi_i^{(n)} \tag{3.23}$$

[10] A method is said "*size consistent*" when the energy computed for a molecular system by bringing two of its subsystems (say A and B) at infinite distance is equal to the sum of that of the two subsystems A and B computed as separate ones.

and

$$E_i = \sum_{n=0} \lambda^{(n)} E_i^{(n)} \tag{3.24}$$

with

$$E_i(n) = \langle \Phi_i^{(n)} | H^{(n)} | \Phi_i^{(n)} \rangle \tag{3.25}$$

and $\lambda^{(0)} = 1$. By inserting above relationships into the Schrödinger equations and properly collecting the terms of a given order one obtains the successive corrections to the wavefunction

$$\Phi_i = \sum_j c_j^{(k)} \Phi_j^{(0)} \tag{3.26}$$

with

$$c_j^k = \frac{1}{E_j^{(0)} - H_{jj}} \sum_i \left(V_{ji} c_i^{(k-1)} - \sum_{n=0}^{k-1} E_j^{(k-n)} c_i^{(n)} \right) \tag{3.27}$$

where

$$V_{ji} = \int \Phi_j^{(0)} V \Phi_i^{(0)} d\tau \tag{3.28}$$

$$H_{ji} = \int \Phi_j^{(0)} H \Phi_i^{(0)} d\tau \tag{3.29}$$

where $d\tau$ represents the integration with respect to all the electronic coordinates. In the case of multireference methods internal (i.e., occupied in the reference configuration) and external (i.e., unoccupied in the reference configuration) molecular orbitals are grouped differently and procedures have been made very efficient especially when considering single and double excitations.

Many-body perturbation methods include electron correlation ensuring both size consistency and size extensivity.[11] In the perturbation theory of Möller–Plesset the unperturbed Hamiltonian is written as a sum of Fock F_j operators defined in Eq. 3.18 with the perturbation $V = H - \sum F$ operator. The eigenfunctions and the eigenvalues of the Fock operators are the molecular orbitals ϕ_j and related energies ϵ_j respectively. By truncating the infinite series (exact solution) to the second and third term one obtains the MP2 and MP3 approximate solution (see [23]). Such procedure gives energies $E^{(0)}$, $E^{(1)}$ and $E^{(2)}$ proportional to the number K of considered electrons showing the *size consistency* of the perturbative level of theory.

[11] A method is said "*size extensive*" when the energy computed for N noninteracting (identical) molecules is equal to N times the energy of a single molecule.

3.3 Toward Extended Applications

3.3.1 Computation of Other Molecular Properties

As you have already seen in Chap. 2, the energies of the orbitals and the coefficients of the expansion of molecular orbitals into atomic orbitals obtained by solving the Schrödinger equation can be used to compute other properties of the investigated system. In other words, one can evaluate the physical observables associated with a given quantum mechanical operator (say $\hat{O}$) to compute its eigenstates

$$\hat{O}\Phi = \omega\Phi \tag{3.30}$$

and average value

$$\omega = \frac{\int \Phi^* \hat{O} \Phi d\tau}{\int \Phi^* \Phi d\tau} = \frac{< \Phi|\hat{O}|\Phi >}{< \Phi|\Phi >}. \tag{3.31}$$

As a consequence, the accurate and detailed investigation of the electronic structure of the considered system is of paramount importance for determining a large variety of the observable properties of chemical processes. As a matter of fact, the search of either the equilibrium geometry of a molecular system or of the structures of the transition states which are characterized by saddle points is greatly helped by the possibility of obtaining from ab initio calculations information on the potential derivatives with respect to the internuclear distances. In order to characterize the stationary points of the potential energy surface one requires the Hessian matrix containing second derivatives. Obviously, for stationary points, the first derivatives of energy with respect to geometry changes is zero. The criterion for a minimum is that all the eigenvalues of the Hessian matrix are positive, while for a saddle point corresponding to a transition state one requires all except one of the eigenvalues of the Hessian to be positive. The procedures consider small displacements d_i from the equilibrium configuration after expressing the potential energy function as a Taylor-series expansion in the displacement coordinates

$$V = \frac{1}{2!}\sum_i\sum_j \left(\frac{\partial^2}{\partial d_i \partial d_j}\right)_{eq} d_i d_j + \frac{1}{3!}\sum_i\sum_j\sum_k \left(\frac{\partial^3}{\partial d_i \partial d_j \partial d_k}\right)_{eq} d_i d_j d_k + ... \tag{3.32}$$

The derivatives of the potential energy with respect to the displacement coordinate d_i evaluated at the equilibrium configuration are the force constants (with the first one missing because being null at stationary points). Thus $\left(\frac{\partial^2}{\partial d_i \partial d_j}\right)_{eq}$ is a harmonic force constant f_{ij} while f_{ijk} and f_{ijkl} are the cubic and quartic force constants associated with third and fourth derivatives, respectively. Of course, derivatives can be evaluated numerically by finite differences. For the gradient, however, this involves several additional calculations and is often affected by large numerical inaccuracy. On the contrary, the corresponding analytical evaluation requires only twice the time of a single SCF calculation. The calculation of analytic second derivatives still requires only three or four times the time taken by the gradient calculation.

This method is used mainly as a test of the accuracy of the calculations and for their improvement. The comparison can be made also with a given property like the vibrational frequencies of the molecular system or a set of properties (multiproperty analysis) owing to the fact that each property may be more sensitive to different features of the potential energy. Among the quantities to consider are structural and thermodynamical properties (conformational maps, free energy maps, ionization potentials, electronic affinities), charge distributions (Mulliken population analysis, molecular potential fields, electronic density at the border), dynamical properties (cross sections, energy distribution of products, vector distributions), transportation properties (viscosity, virial).

To this end, it is important to point out here that the analytical calculation of the derivatives of the potential energy can be significantly helpful in determining dipole moment and polarizability derivatives. Derivatives of the potential energy also provide invaluable information for the fitting of the calculated potential energy values to a functional form describing the path connecting reactants and products as we shall discuss later.

3.3.2 *Density Functional Theory Methods*

For large systems and a large number of molecular geometries calculations (as often needed for the investigation of molecular collisions) electronic density funtional theory (DFT) provides a versatile and practical means to calculate electronic energies because there is a linearly scaling with the number of electrons (while the HF methods scale usually as the fourth power). The key quantity of DFT is the electronic density $\rho(\mathbf{r})$ (the number of electrons per unit volume in a given space point). The method replaces, in fact, the problem of determining the wavefunction (that is the key quantity of traditional ab initio techniques) with that of determining the electronic density. The method leverages on the simple idea that the electronic system of a molecule behaves as a gas of particles subject to coulomb interactions.

The electronic density $\rho(\mathbf{r})$ depends on three spatial coordinates and on the spin independently of the dimension of the actual physical system. This has been formalized by the Hohenberg–Kohn theorems stating that for the ground state the potential energy $V(\mathbf{r})$ depends on the electronic density $\rho(\mathbf{r})$ that can be computed using variational methods. At the same time $\rho(\mathbf{r})$ can be easily related to the number K of electrons as follows:

$$\int \rho(\mathbf{r})\mathrm{d}\mathbf{r} = K \tag{3.33}$$

and determines the wavefunction of the ground state as well as the related electronic energy that is formulated as

$$V(\mathbf{r}) = T_e[\rho] + V_{ext}[\rho] + V_{ee}[\rho] \tag{3.34}$$

where $T[\rho]$ is the kinetic energy, $V_{ext}[\rho]$ the electron-nucleus attractive energy and V_{ee} the electron–electron repulsive energy. In turn, V_{ee} can be expressed as the sum of a term of classical repulsion $J[\rho]$ (Coulomb potential) and a nonclassical term $E_{exc}[\rho]$ containing electronic correlation

$$V_{ee}[\rho] = J[\rho] + E_{exc}[\rho]. \tag{3.35}$$

The computational scheme of DFT is embodied into the Kohn–Sham equations and the effect of the interaction among K electrons is rendered as that of the same number of noninteracting electrons though subject to an external potential. Accordingly, the total energy can be written as:

$$E[\rho] = T_s[\rho] + V_{ne}[\rho] + J[\rho] + E_{exc}[\rho] \tag{3.36}$$

where $T_s[\rho]$ is the energy associated with a gas of noninteracting electrons ($T_s[\rho] = \sum_i^N \langle\psi_i| - \frac{1}{2}\nabla^2|\psi_i\rangle$), V_{ne} is the interaction energy with the external potential $V(r)$ with density $\rho(r)$

$$V_{ne} = \int \rho(r)V(r)\mathrm{d}r \tag{3.37}$$

The last term of Eq. 3.36 E_{exc} is the so-called exchange and correlation energy and contains the differences between the independent electron gas model and the real system including the nonclassical part of V_{ee}. By applying the variational principle to the energy formulation one gets K Kohn–Sham monoelectronic equations:

$$\left[-\frac{\hbar^2}{2m}\nabla^2 + V_{eff}(r)\right]\psi_i = \epsilon_i\psi_i \tag{3.38}$$

Due to the fact that $V(\mathbf{r})$ depends on $\rho(r)$, related equations are solved using an iterative SCF-like procedure. Like in the Hartree–Fock theory the many-electron problem leads to the resolution of K monoelectronic equations. However, while in the HF theory the electronic correlation effects are introduced either through multiconfigurational methods or by expanding the many-electron wavefunction in Slater determinants, the DFT theory directly incorporates the effect of the electronic correlation.

3.3.3 *The Valence Electron Method*

Although codes running ab initio calculations have progressed enormously in terms of speed and efficiency, the growing attention for large molecular systems has prompted the use of more approximate methods. For example, the fact that the number of bi-electron integrals to be computed goes as the fourth power of the dimension of the basis function adopted, has led to the development of methods cutting down on it. For example, by taking into account that internal electrons (core) play a minor role in determining chemical properties, the Hamiltonian can be decomposed as follows:

$$H = H^{core} + \sum_j J_{jj} - \sum_j K_{jj} \tag{3.39}$$

where H^{core} embodies the kinetic energy and the interaction of the electrons of the internal shells.

The matrix elements containing H^{core} are usually replaced either by theoretical or empirical quantities. In particular, if the molecule contains a heavy atom (like in the case of metal complexes), this may be rendered by adopting a pseudopotential or an effective potential leading to a significant saving of computing time.

For this one can adopt two different strategies with respect to the use of parameters in molecular orbital calculations. The first strategy moves from the consideration that ab initio calculations are themselves an approximation and that corrective parameters are introduced in any case to force agreement with the experiment. The other strategy considers (as done already when choosing to adopt STO-NG rather than pure STO orbitals) the separate calculation of certain quantities as rigorously as possible and the utilization of their outcomes whenever possible.

3.3.4 Dropping Multicenter Integrals

A seemingly drastic simplification (Neglected Differential Overlap, NDO) is the dropping of multicenter integrals

$$\int \chi_m(1)\chi_n(1)\frac{1}{r_{12}}\chi_j(2)\chi_l(2)d\tau_1 d\tau_2 \tag{3.40}$$

by assigning them the value of a Kronecker δ function. Multicenter integrals are difficult to evaluate when the atomic functions are centered on different atoms. To this end the $(mn|jl)$ integral can be written as

$$(mn|jl) = \delta_{mn}\delta_{jl}(mm|jj) \tag{3.41}$$

where

$$(mm|jj) = \int \chi_m(1)\chi_m(1)\frac{1}{r_{12}}\chi_j(2)\chi_j(2)d\tau_1 d\tau_2. \tag{3.42}$$

This makes equal to zero the integrals concerned with 3 and 4 centers as well as several one and two centers integrals when the orbitals considered for one of the two electrons of interest are different. Even more drastic is the CNDO (Completely

Neglected Differential Overlap) that sets equal to zero all overlap integrals but those between valence orbitals. Moreover, also the integrals between valence integrals are approximated either to an (arbitrary) reference value or to experimental data like the ionization potential. A mitigation criterion that connects to the value of the exchange integral K to the difference in energy of opposed and parallel spin integrals is the one named INDO. In this case, integrals of the type $(ml|ml)$ are not neglected if the functions χ_m and χ_i are centered on the same atom. However, the role played by these approaches has to be understood in terms of the possibility of covering large regions of the molecular geometries to be considered in dynamical processes. In this case, in fact, the approximate methods discussed here are mainly used to best fit locally the potential energy values to a suitable functional representation by adjusting the value of related parameters.

3.4 Full Range Process Potentials

3.4.1 The Three-Body Internuclear Coordinates

When tackling the problem of describing reactive processes at atomistic level one needs to calculate the electronic structure of completely different molecular arrangements including those far from the equilibrium geometry. In order to better illustrate this case, we concentrate here on the simplest reaction prototype that consists of three atoms. Three-atom systems are, in fact, the ideal case study both because they include the most investigated prototype atom–diatom reactive and nonreactive processes (that are commonly used to the end of rationalizing chemical reaction mechanisms) and because they are on the theoretical side the simplest reactive case to handle and to understand. In order to describe chemical reactive processes, time t is the ideal continuity variable. However, once that time (that is an ideal continuity variable for describing chemical reactions) has been factored out, as it happens in time independent approaches, one faces the problem of devising a suitable alternative continuity variable out of the position coordinates.

As already done for two-body systems, the first step is the reduction of the number of position vectors by separating the centre-of-mass (CM) and the related motion. From the vectors $\mathbf{W}_A$, $\mathbf{W}_B$, and $\mathbf{W}_C$ (the position vectors of the nuclei in the chosen axis frame omitted in the picture for the sake of clarity) we can build the three internuclear vectors $\mathbf{r}$ (see Fig. 3.1) using the relationships (please notice hereinafter the change of meaning of $\mathbf{r}_{AB}$, $\mathbf{r}_{BC}$ and $\mathbf{r}_{CA}$ from an electron-nucleus vector to a nucleus–nucleus one)[12]

$$\mathbf{r}_\nu = \mathbf{r}_{\lambda\mu} = \mathbf{W}_\lambda - \mathbf{W}_\mu \tag{3.43}$$

[12]Single subscript notation is used mainly in the scheme of the separated atom formalism (in which the label singles out the unbound atom) while the double subscript notation is mainly meant to single out the bound atoms.

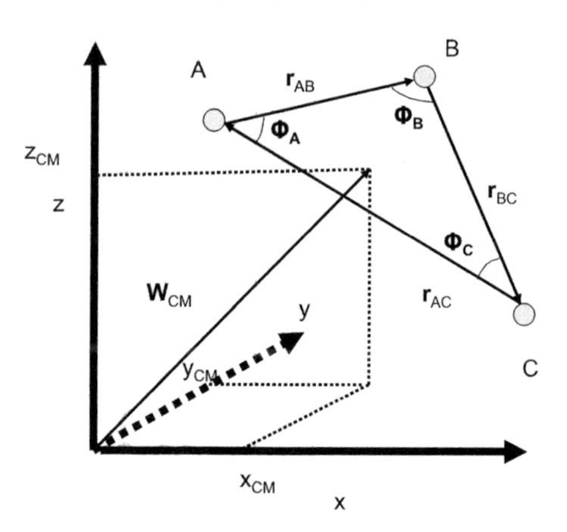

Fig. 3.1 Internuclear distances of a three-body system and related angles

(where λ, μ, and ν are a cyclic permutation of the elements of the sequence A, B, and C with τ being the label of the generic arrangement) and the related internuclear distances ($|\mathbf{r}_{\lambda\mu}|$). Internuclear distances are a very popular set of internal coordinates widely used to describe the interaction of the system. As done already for the two-body problem, one can adopt a body-fixed reference frame by orienting the cartesian coordinates so as to have an axis jacent in the plane defined by the three bodies and pointing on a specific direction (say for example the incoming or the outgoing atom) obtained through a proper rotation of the reference frame. The internuclear distances form in pairs with the angles Φ_τ (also labeled sometimes after the isolated or exchanged atom) and obey the triangular rule (namely one internuclear distance can neither be larger than the sum of the other two nor be smaller than their difference). Obviously, for the same geometric reason one can use two distances and the included angle or one distance and the two adjacent angles. The set of internuclear distance coordinates has the undoubtable advantage of being well suited for representing and formulating (as we shall discuss later) two and more body interactions at the same time. They are, in fact, well suited to describe both the strong and intermediate range as well as the isolated atom, diatom, and polyatom geometries. As a matter of fact, the most popular procedures for calculating and representing (both analytically and graphically) the potential energy also for large polyatomic systems make use of fine-grained grids of values of internuclear distances.

3.4.2 Global Formulation of the Potential Energy Surface

Early reactive scattering studies were almost exclusively based on the use of the LEPS [24] model potential. As a matter of fact, the LEPS PESs has been systematically used to rationalize the behaviour of atom–diatom reactions (see some instructive examples at the end of the next chapter). As shown by its functional formulation the LEPS was derived for the family of atom–diatom systems by an oversimplified ab initio treatment of $H + H_2$. In the notation of Eq. 3.43 the LEPS reads as

$$V(r_\tau, r_{\tau+1}, r_{\tau+2}) = \sum_{\tau=1}^{3} J(r_\tau) - \frac{1}{2}\sqrt{\sum_{\tau=1}^{2}\sum_{\tau'>\tau}^{3}[K(r_\tau) - K(r_{\tau'})]^2} \tag{3.44}$$

in which both the J (coulomb) and K (exchange) terms are formulated as a combination (weighted by the Sato parameters, one for each pair of atoms, which are a reminiscence of the overlap integral) of the Morse ($D_\tau n_\tau(n_\tau - 2)$) (already illustrated in Chap. 2) and the anti-Morse ($D_\tau n_\tau(n_\tau + 2)/2$) potentials. Accordingly, the LEPS has been always considered as an empirical functional form whose parameters (the Sato parameter and its angular dependence, if any) are varied to the end of optimizing the reproduction of theoretical and/or experimental data using a weighted Least Square (LS) method [25]. This feature will turn out to be useful when trying to extend the versatility of the BO variable ($n = e^{-\beta(r-r_0)}$).

The scarce flexibility and the difficult extensibility of the LEPS, however, have prompted the formulation of other global functional representations for atom–diatom PESs. A popular global formulation of the reactive PES was born out of the generalization of a properly damped polynomial P^{MR} (of an arbitrary degree and possibly of the appropriate symmetry) in the related internuclear distances quenched at long range by an exponential damping function [26] whose parameters are LS best fitted to accurate ab initio potential energy values

$$V(\mathbf{r}) = V(r_\tau, r_{\tau+1}, r_{\tau+2}) = P^{MR}(r_\tau, r_{\tau+1}, r_{\tau+2}) n_\tau n_{\tau+1} n_{\tau+2}. \tag{3.45}$$

A weakness of this formulation is the possible formation of spurious structures in the intermediate range due to dominance of the divergence of the polynomial term over the damping effect of the exponential factor as the internuclear distances increase.

A more appropriate LS alternative global formulation of the PES is a polynomial in the BO variables (P^{BO}), again of an arbitrary degree and possibly of the appropriate symmetry thanks to their built-in proper behaviour at long range [4, 27]

$$V(\mathbf{r}) = V(r_\tau, r_{\tau+1}, r_{\tau+2}) = P^{BO}(n_\tau, n_{\tau+1}, n_{\tau+2}). \tag{3.46}$$

Other global analytical representations of the PES have also been formulated in terms of products of BO and internuclear distances [28].

The most general procedure for formulating a global PES is based on the weighted least squares (LS) method [25] whose formalism is closely followed here. The LS method expands the PES in terms of the $\mathbf{f}_k(\mathbf{r})$ basis functions depending on the collection of coordinates $\mathbf{r}$ on which it is formulated with c_k being the coefficients of such expansion

$$V(\mathbf{r}) = \mathbf{c}^T\mathbf{f}(\mathbf{r}) = \mathbf{f}^T(\mathbf{r})\mathbf{c} = \sum_{l=1}^{L} c_l f_l(\mathbf{r}). \tag{3.47}$$

In Eq. 3.47 $\mathbf{c}$ and $\mathbf{f}$ are column vectors, L is the number of basis functions, $\mathbf{r}(i)$ and $v(i)$ are the coordinates and energy values of the data points to be interpolated and the superscript T denotes as usual "transpose". In a LS method one minimizes the functional of the sum of the weighted squares of the deviations of the fitted potential from the calculated data to determine the coefficients c_k (the weights, often taken to be unity, can be sometimes chosen to weight more the points located around the minimum energy path of the considered process channels).

Functions f_k can be freely chosen. In fact, for what we have already discussed at the beginning of the previous section about the evaluation of molecular properties, regardless of the procedure adopted for the ab initio calculations the resulting potential energy values can be traced back to a set of basis functions which need only to be flexible enough to properly reproduce their main features. The most popular choices are the already above considered polynomials in internuclear distances and exponentials (including mixed ones) ensuring a correct behaviour at the asymptotes. Another important feature is the smoothness of the fitting avoiding spurious structures in localized regions of the potential. In this respect the use of polynomials in BO coordinates (thanks to their intrinsic vanishing at long distance and divergence a short ones [4, 27]) is safer. The enforcement of the symmetry of the system on the formulation of the PES can also be adopted to the end of reducing the number of terms [29].

3.4.3 Local and Mobile Methods

More recently, the increasing availability of ab initio estimates of the potential energy values for an increasing very large number of molecular geometries has fostered the use of local methods. A great advantage of these methods is the fact that the fitting can be improved (if looking for new geometries or unsatisfied with the available fit) by simply adding more for nearby ab initio points. Moreover, the points need not be located on a uniform grid.

A popular local method is the Shepard one in which the potential energy surface $V(\mathbf{r})$ is represented by a weighted sum of Taylor expansions $T_i(\mathbf{r})$ about each ab initio point:

$$V(\mathbf{r}) = \sum_{i=1}^{i_{max}} w_i(\mathbf{r})T_i(\mathbf{r}), \tag{3.48}$$

where $\mathbf{r}$ is a vector of $3N - 6$ internal coordinates, N is, as usual, the number of nuclei and i_{max} is the number of ab initio points. This is based on the assumption that the set of ab initio data is so dense that any geometry of interest belongs to the domain of convergence of the Taylor expansions around at least one ab initio point. The w_i weight functions are now chosen so as to switch on whenever the ab initio points are reasonably close to the geometry being considered and the potential to be given by the weighted average of Taylor expansion estimates from the nearby points. To the end of smoothly interpolating between adjacent ab initio points, Collins and coworkers [30] choose as internal coordinates $\mathbf{z} = 1/\mathbf{r}$ so that a single ab initio point used in Eq. (3.48) describes the asymptotic behavior of the isolated diatomic potential quite accurately (Taylor expansions in inverse coordinates have a much larger domain of convergence than the coordinates themselves).

The w_i weight functions are formulated as inverse powers p of the sum of all the $(z_k - z_k(i))^2 + k^2$ terms with the power parameter determining the drop off of the weight function and the parameter k avoiding singularities while ensuring sufficient sharpness near each data point.

An interesting evolution of the local methods is represented by the local mobile (LM) LS ones in the local basis functions are modulated as a function of the molecular geometry of interest [25]. Because of this, lower order polynomial functions are needed though the coefficients of the basis functions are now varying with the geometry and set a heavier computational demand. As a matter of fact in the LM-LS scheme [31] the value V at point $\mathbf{r}$ is represented by a linear combination of linearly independent basis functions $f_k(\mathbf{r})(j = 1, \ldots, n)$, as follows:

$$V(\mathbf{r}) = \mathbf{c}^T(\mathbf{r})\mathbf{f}(\mathbf{r}) = \mathbf{f}^T(\mathbf{r})\mathbf{c}^T(\mathbf{r}) = \sum_{l=1}^{L} c_l(\mathbf{r}) f_l(\mathbf{r}), \tag{3.49}$$

where the coefficients $c_1(\mathbf{r}), c_2(\mathbf{r}), \ldots, c_L(\mathbf{r})$ depend on the coordinates $\mathbf{r}$.

Being as before the coordinates and energy values to be interpolated $\mathbf{r}(i)$ and $v(i)(i = 1, 2, \ldots, i_{max})$ with i_{max} the number of data points, the error functional is formulated as

$$\sum_{i=1}^{i_{max}} w_i(\mathbf{r})[V(\mathbf{r}) - v(i)]^2. \tag{3.50}$$

This provides fresh ground for the use of previously proposed formulations of the PESs. Among them is the Diatomic In Molecule (DIM) [32] method that is a simple method to deal with theoretical studies of electronically nonadiabatic transitions. The LM-LS scheme fuels also new interest in the formulation of the PES in terms of many-body expansions (MBE) defined as follows:

$$V(\mathbf{r}) = \sum_{pairs} V^{(2)}(\mathbf{r}^{(2)}) + \sum_{triples} V^{(3)}(\mathbf{r}^{(3)}) + \cdots + \sum_{Nuples} V^{(N)}(\mathbf{r}^{(N)}). \tag{3.51}$$

and that allows to modulate the various each many-body component as the process progresses. In particular, for a three-body system this means three two-body terms and one three-body term of the type

$$V(\mathbf{r}) = V^{(2)}(r_{1-2}) + V^{(2)}(r_{2-3}) + V^{(2)}(r_{3-1}) + V^{(3)}(r_{1-2}, r_{2-3}, r_{3-1}) \quad (3.52)$$

while for a four-atom system this becomes six two-body terms, four three-body terms, and one four-body term.

For example, it provides further motivations to the separate evolution of the three or more body terms of the double MBE that are partitioned in a first term accounting for the Hartree–Fock contribution and a second term accounting for dynamic correlation contributions to the interaction [33] despite the fact that the separation of the different components is neither obvious nor unique.

3.4.4 Process-Driven Local and Mobile Fitting Methods

A strengthening of the local mobile fitting methods can be obtained by embodying in the procedure a criterion for guiding the selection of **c** and **f** via the relevance of the considered process to the so-called many-process expansion (MPE) [34] and the possibility of leveraging on a flexible continuity variable driving the switch from one molecular arrangement to another. In this respect, the Bond order coordinates turn out to be particularly useful because of their correct behavior at both ends and of the confinement of the interaction into a finite space (see, for example, the comparison of the representation of the Morse potential in internuclear distance and in the BO variable given in Fig. 3.2). In the BO space, the Morse potential has an inverse and truncated Harmonic-like shape equal to zero at $n = 0$ and a minimum at $n = 1$. This inverted nature of the BO space with respect to the physical one allows also a proper formulation of the atom–diatom long range interactions using polynomials in the related variables [4]. Accordingly, the B exchange process A + BC → AB + C (as will be discussed in more detail below one can also consider the C exchange process B + CA → BC + A and the A exchange process C + AB → CA + B) can be formulated in terms of diatomic-like ROBO potentials rotating around the common origin of the two involved BO variables [35]. The related rotation angle α defined as

$$\alpha = \arctan\left[\frac{n_{AB}}{n_{BC}}\right] \quad (3.53)$$

that is a continuity variable in the B transfer process transforming the reactant diatom BC into the related product AB. At the same time the variable ρ_B defined as:

$$\rho_B = [n_{AB}^2 + n_{BC}^2]^{1/2} \quad (3.54)$$

spans the different fixed angle elongations of the system. The corresponding fixed arrangement angle Φ_B ROBO potential channel(s), can be formulated as a polynomial in ρ_B as follows:

$$V_B^{BO}(\Phi_B; \alpha, \rho_B) = D(\Phi_B; \alpha)P(\Phi_B, \alpha; \rho_B) \quad (3.55)$$

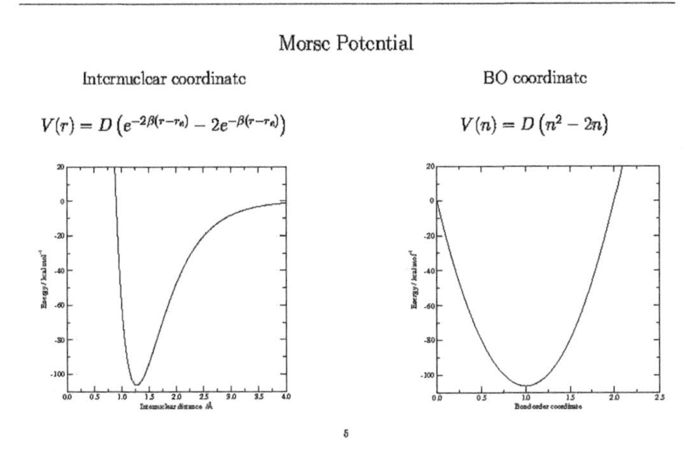

Fig. 3.2 Morse potential represented in the physical space (lhs panel) and in the BO space (rhs panel) where $n = e^{-\beta(r-r_e)}$

in which $D(\Phi_B; \alpha_B) = \sum_{j=0}^{J}[d_{s_j}\sin(j\alpha) + d_{c_j}\cos(j\alpha)]$ describes the evolution of the fixed Φ_B minimum energy of the potential energy channel from reactants (at $\alpha = 0$) to products (at $\alpha = \pi/2$) and the polynomial $P(\Phi_B, \alpha; \rho_B)$ describes the shape of the B channel cut while the system elongates or contracts out of its (fixed Φ_B) minimum energy geometry. The mentioned characteristics of α make it a variable of election for driving not only the formulation of the interaction but also its fitting in a process driven fashion. In the particular case of $N + N_2$ discussed in Ref. [36] the following simple formulation

$$D(\Phi_N; \alpha) = -D_e + S_B(\Phi_B)\sin(2\alpha) \tag{3.56}$$

was adopted to the end of fitting the single barrier LEPS thanks both to the collinearity of the transition state (TS) and to the symmetry of the system. In this case S_B is equal to the value of the potential energy of the collinear saddle E^{TS} and increases when moving away from the collinear arrangement according to a relationship of the type

$$S_B = \sum_{k=1}^{kmax} E^{TS}(\Phi_{Bk}^{TS} - \Phi_B)^{2(k-1)}. \tag{3.57}$$

However, by playing with the flexibility of this simple formulation of the LAGROBO model it was possible to easily modify the structure of the PES from collinear to bent as shown in Table 3.1 where the transition state features of the reaction channel of both the $N + N_2$ LEPS and two LAGROBO PESs (the origin of such

Table 3.1 Comparison of the features of the reaction channel of the LEPS with those of two LAGROBO PESs

PES	$r_1 = r_2/a_o$	$\alpha/^o$	E^{TS}/eV
LEPS	2.34	180	1.55
LAGROBO3	2.37	125	1.40
LAGROBO4	2.24	117	2.06

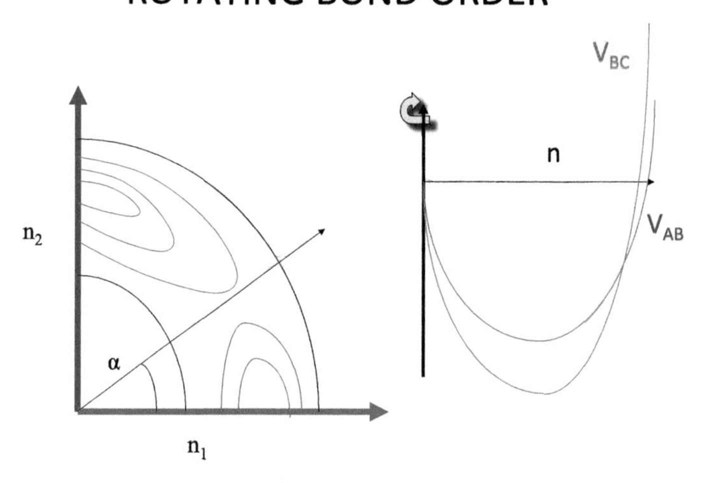

Fig. 3.3 LHS panel: contour plots of the BO potential for the reaction A + BC → AB + C represented in the BO space as a ROtating Bond Order (ROBO) as a function of the BO variables (RHS panel) where $n_i = e^{-\beta_i(r_i - r_{ei})}$; RHS panel: evolution of the ROBO cut in going from reactants to products while rotating from $\alpha = 0$ to $\alpha = \pi/2$

name is given below) are shown. The same approach has been used to introduce a well on top of the transition state barrier [36–38].

This assigns to the angle α of Fig. 3.3 (where the label 1 corresponds to the reactant diatom BC and label 2 corresponds to the product diatom AB of the mentioned process A + BC → AB + C) the role of a continuity variable of the reactive process. The BO potential, in fact, by rotating around the zero of the BO variables in the ROtating BO (ROBO) [35] model potential smoothly connects in the chosen arrangement reactants to products. Moreover, in the MPE spirit the BO formalism allows a straightforward switch from the A + BC → AB + C process to the already mentioned B + CA → BC + A and C + AB → CA + B ones using a weight depending on the closeness of the arrangement angle Φ to a reference angle (the choice of the largest angle (i.e., the preference for the most collinear configuration) has motivated the adoption of the LAGROBO acronym) [39]. Further advantage can be taken also by the adoption of the space reduced (SRBO) formulation of the BO variables [40] that allows a wise sampling of the interaction space to the end of bal-

Table 3.2 Coefficients of the N_2 BO potentials formulated using $D_e = 954.92$ kJ/mol and $r_e = 0.1098$ nm. RMSD(BO4) = 4.33 kJ/mol, RMSD(BO6) = 2.72 kJ/mol

PES	c_1	c_2	c_3	c_4	c_5	c_6
BO4	2.4200	−1.9573	0.6547	−0.1174		
BO6	2.9833	−3.7743	2.9145	−1.4858	0.4077	−0.0457

ancing the representation density of both the short and long distance geometries as well as its specialization on the desired regions of the interaction. All this will turn out to be useful when generalizing the LAGROBO model to more than three bodies (see Chap. 5).

An additional key feature of the BO formulation of the interaction is associated with the fact that it can quite naturally incorporate the fitting to the long range contributions. This can be obtained through the use of a higher order single BO polynomial globally fitting both the long and the short range interaction. Values of the BO polynomial coefficients and of the root mean square deviation obtained for the N_2 case [4] are shown in Table 3.2 (upper row for the fourth order and lower row for the sixth order).

3.5 Problems

3.5.1 *Qualitative Problems*

1. **Atomic Structure**: Describe an electronic structure calculation for the He and Ne atoms using atomic orbitals.
2. **Electronic Structure**: Describe a electronic structure calculation of HF using molecular orbitals. Draw an electron correlation diagram for H_2, Li_2, Be_2, and HF.
3. **Hartree–Fock**: Explain why the Hartree–Fock method will only give approximate results. Explain what one is trying to achieve when one uses a CI or MC-SCF.
4. **CISD**: Clearly explain what is meant by single and double configurations using the Be atom as an example.
5. **He**: What is the ground state energy of the He atom if one assumes the electrons are noninteracting. Compare this estimate to the correct ground state of this atom.
6. **Overlap Integrals**: Explain why the eigenvalues of the overlap matrix are greater than zero and give a physical meaning for the eigenvalues of the overlap matrix.
7. **Fine and Hyperfine Structure**: The electrons have an intrinsic spin as well as a spin angular momentum, most nuclei also have a nonzero nuclear spin. Explain why you think these properties might have an effect on the energy levels and spectrum. Read the scientific American article on the hydrogen atom [41].

3.5.2 Quantitative Problems

1. **2D Monte Carlo**: Develop a Monte Carlo code to calculate the numerical value of π. To do this consider a square box centered at the origin with a dimension of 2 units. A circle of radius $r = 1$ also centered at the origin will fit just inside the box. The area of the box and the circle are 4 and π, respectively. For each iteration use a random number generator to obtain x and y values in the range -1 to 1. If $r^2 = x^2 + y^2 < 1$ the point is within the circle otherwise it is inside the box but outside the circle. Print the ratio of the number of point inside the circle divided by the number of points outside the circle after 1,000, 10,000, 100,000, 1,000,000 iterations. This ratio should converge to $\pi/4$ which is just the ratio of the areas. The Monte Carlo is not very efficient for doing iterated integrals with less than 7–10 dimensions so don't expect a rapid convergence. See the next problem to understand why the Monte Carlo method is so useful in many areas of science.
2. **Multidimensional Monte Carlo**: Now modify your code to calculate the 10-dimensional integral:

$$I = \int_{-1}^{1} dx_1 \int_{-1}^{1} dx_2 \int_{-1}^{1} dx_3 \cdots \int_{-1}^{1} dx_{10} \left[x_1^2 + x_1^2 + x_1^2 + x_1^2 \cdots x_{10}^2\right] \quad (3.58)$$

   If you attempt to perform this integral using the trapezoidal or similar griding techniques with 10 point for each coordinate you will need 10^{10} points! This clearly demonstrates why a Monte Carlo method is efficient for integrals with many dimension as is done in the quantum Monte Carlo method.
3. **He Atom**: Use two basis functions e^{-2r} and e^{-4r} in the Hartree–Fock method to determine the approximate ground state energy of the He atom. Can you find exponential parameters better than -2 and -4?
4. **He Atom - again**: Use the numerical Hartree–Fock code supplied in the additional material to calculate the ground state energy of He. This code will give the numerically accurate answer for the Hartree–Fock ground state energies. Carefully explain why this answer is not the experimental value which is also extremely close to an accurate theoretical calculation. You might want to try Robert D. Cowans atomic structure code https://www.tcd.ie/Physics/people/Cormac.McGuinness/Cowan/ to get additional accuracy, information, and physical properties about He and other atoms.
5. **Molecular Orbitals**: Using the Gamess molecular structure code calculate the collinear ground state potential energy curves for H_2 and HF.
6. **Molecular Orbitals**: Using the Gamess molecular structure code calculate the ground state potential energy curves for H_3 and Li_3 at several internuclear distances. How do you answers compare with other theoretical results?
7. **Least squares fitting**: Use the linear least squares procedure to determine the C_6 and C_8 coefficients in the long range van der Waals expansion for the Li_2 dimer. First set all of the weights equal to 1 and then increase the weights for larger r values (Table 3.3).

Table 3.3 The radial distance in a_0 and the calculated potential energy in Hartrees

r	V(r)	r	V(r)
6.0	−0.201455806	11.0	−0.001459946
7.0	−0.052413419	12.0	−0.000779729
8.0	−0.017153729	13.0	−0.000444424
9.0	−0.006674236	14.0	−0.000267073
10.0	−0.002964858	15.0	−0.000167664

$$V(r) = -\frac{C_6}{r^6} - \frac{C_8}{r^8} \tag{3.59}$$

Chapter 4
The Treatment of Few-Body Reactions

This chapter focuses on the problem of determining the reactive dynamics of the simplest prototypes of elementary chemical reactions starting from a general non-Born–Oppenheimer (mixed electron–nuclei) approach first and then formulating the problem using a separating from that of the nuclei. To this end, the problem of adopting coordinate sets suited for describing both the interaction and the dynamics of the simplest reactive systems is discussed. Typical features of the atomistic phenomenology of atom–diatom systems such as the effect of a different allocation of energy to the various degrees of freedom in promoting reactivity, the importance of providing an accurate representation of the potential energy, the merits and demerits of reduced dimensionality calculations, and the importance of periodic orbits are analyzed.

4.1 The Combined Dynamics of Electrons and Nuclei

4.1.1 The N-Body Dynamical Equations

In the previous chapter, we have discussed the numerical integration of the poly-electronic multidimensional Schrödinger equation. During a chemical process, the faster electronic motion rapidly adjusts to the sluggish movement of the heavy nuclei and the two motions interplay within a common game. Accordingly, in general, for a system made of N nuclei (each of mass M_i and charge $Z_i q_e$), K electrons (with mass m_e and charge $-q_e$), localizable in space (with respect to an arbitrary axis systems) using the nuclei position vector $\mathbf{W}^{\ddagger}$, and the electron position vector $\mathbf{w}^{\ddagger}$ (the nuclei position vector is of length N $\mathbf{W}^{\ddagger} = (\mathbf{W}_1^{\ddagger}, \ldots, \mathbf{W}_N^{\ddagger})$ and the electron position vector is of length K $\mathbf{w}^{\ddagger} = (\mathbf{w}^{\ddagger}{}_1, \ldots, \mathbf{w}_K^{\ddagger})$), the equation of Schrödinger in its general time-dependent form is

A. Laganà and G. A. Parker (eds.), *Chemical Reactions*, Theoretical Chemistry and Computational Modelling, https://doi.org/10.1007/978-3-319-62356-6_4

$$i\hbar\frac{\partial}{\partial t}\Psi(\mathbf{W}^{\ddagger},\mathbf{w}^{\ddagger},t)=\mathcal{H}_{tot}(t)\Psi(\mathbf{W}^{\ddagger},\mathbf{w}^{\ddagger},t), \tag{4.1}$$

where $\mathcal{H}_{tot}(t)$ is the total Hamiltonian operator (that is independent of time if the system is conservative) and $\Psi(\mathbf{W}^{\ddagger},\mathbf{w}^{\ddagger},t)$ is the time-dependent wavefunction (it is worth noting here that nuclei position vectors now are not parameters). The total Hamiltonian $\mathcal{H}_{tot}(t)$ is the sum of the operators nuclear kinetic energy ($\mathcal{T}_n$), electron kinetic energy ($\mathcal{T}_e$), and three potential energy terms: nucleus–nucleus($\mathcal{V}_{nn}$), electron–nucleus ($\mathcal{V}_{ne}$), and electron–electron ($\mathcal{V}_{ee}$)

$$\mathcal{H}_{tot}=\mathcal{T}_n(\mathbf{W}^{\ddagger})+\mathcal{T}_e(\mathbf{w}^{\ddagger})+\mathcal{V}_{nn}(\mathbf{W}^{\ddagger})+\mathcal{V}_{ne}(\mathbf{W}^{\ddagger},\mathbf{w}^{\ddagger})+\mathcal{V}_{ee}(\mathbf{w}^{\ddagger}) \tag{4.2}$$

with

$$\mathcal{T}_n=-\sum_{i=1}^{N}\frac{1}{2M_i}\nabla^2_{\mathbf{W}_i^{\ddagger}}\quad \mathcal{T}_e=-\sum_{j=1}^{K}\frac{1}{2m_e}\nabla^2_{\mathbf{w}_j^{\ddagger}} \tag{4.3}$$

$$\mathcal{V}_{nn}=\sum_{i=1}^{N-1}\sum_{i'>i}^{N}\frac{Z_iZ_{i'}q_e^2}{\left|\mathbf{W}_i^{\ddagger}-\mathbf{W}_{i'}^{\ddagger}\right|}\quad \mathcal{V}_{ee}=\sum_{j=1}^{K-1}\sum_{j'>j}^{K}\frac{q_e^2}{\left|\mathbf{w}_j^{\ddagger}-\mathbf{w}_{j'}^{\ddagger}\right|}$$

$$\mathcal{V}_{ne}=-\sum_{i=1}^{N}\sum_{j=1}^{K}\frac{Z_iq_e^2}{\left|\mathbf{W}_i^{\ddagger}-\mathbf{w}^{\ddagger}{}_j\right|}. \tag{4.4}$$

The first simplification will be to use center-of-mass (CM) coordinates for which the position vector of the CM is

$$\begin{aligned}\mathbf{W}_{CM}&=\tfrac{1}{\mathcal{M}}\left(\textstyle\sum_{i=1}^{N}M_i\mathbf{W}_i^{\ddagger}+\sum_{j=1}^{K}m_e\mathbf{w}_j^{\ddagger}\right)\\&\approx\tfrac{1}{M_o}\textstyle\sum_{i=1}^{N}M_i\mathbf{W}_i^{\ddagger},\end{aligned} \tag{4.5}$$

where $\mathcal{M}$ is the total mass of the system and M_o is the total mass of the nuclei (this is actually an approximation to the total center-of-mass because we are neglecting the mass of the electrons. Yet, this is a good approximation because the electron mass is $\approx$ 1822 times smaller than that of the protons and neutrons that constitute the nuclei. The coordinates of the electrons ($\mathbf{w}$) relative to the CM of the nuclei ($\mathbf{W}$) become, therefore,

$$\mathbf{w}_k=\mathbf{w}_k^{\ddagger}-\mathbf{W}_{CM}$$

while those of the nuclei become[1]

[1]Note that for the generic core $\bar{i}$, the following relation holds

$$\mathbf{W}_i = \mathbf{W}_i^{\ddagger} - \mathbf{W}_{CM}.$$

The Hamiltonian operator is thus of the form

$$\begin{aligned}\mathcal{H}_{tot} = &-\tfrac{\hbar^2}{2}\sum_{i=1\neq i\ddagger}^{N}\tfrac{1}{M_i}\nabla^2_{\mathbf{W}_i} - \tfrac{\hbar^2}{2m_e}\sum_{j=1}^{K}\nabla^2_{\mathbf{w}_j}\\ &-\sum_{i=1}^{N-1}\sum_{i'>i}^{N}\frac{Z_i Z_i' q_e^2}{r_{ii'}} - \sum_{j=1}^{K-1}\sum_{j'>j}^{K}\frac{q_e^2}{r_{j'}}\\ &-\sum_{i=1}^{N}\sum_{j=1}^{K}\frac{Z_i q_e^2}{r_{ij}},\end{aligned} \quad (4.6)$$

where $r_{ii'} = |\mathbf{W}_i - \mathbf{W}_{i'}|$, $r_{jj'} = |\mathbf{w}_j - \mathbf{w}_{j'}|$, and $r_{ij} = |\mathbf{W}_i - \mathbf{w}_j|$.

4.1.2 A Direct Integration of the General Equations

In addition to applying the Monte Carlo method (already illustrated in the previous chapter to perform electronic structure calculations) to the integration of the combined electron–nuclei joint motion, various schemes have been proposed for the purpose of coupling nuclear and electronic motions embodied in Eq. (4.1). Typically such effects are taken into account either in terms of derivatives coupling (for adiabatic surfaces) or nonadiabatic potential energy components (for diabatic surfaces) [42]. Other methods try to handle electrons and nuclei on the same footing as in the case of coupled cluster methods [43].

We consider here for simplicity the multiconfiguration time-dependent Hartree (MCTDH) [44, 45] applied to the six-dimensional system of one proton (with position vector $\mathbf{W}$) and one electron (with position vector $\mathbf{w}$) confined within a cavity whose impenetrable walls act as an external force field. In this method, the overall wavefunction is expanded as a sum of configurations with each configuration being a product of single degree of freedom (DOF) wavefunctions (or orbitals). Accordingly, the wavefunction is written as

$$\psi(\mathbf{W}, \mathbf{w}) = \sum_c A_c(t)\Pi_n \phi_{c,n}(z_n, t) \quad (4.7)$$

with z being a suitable coordinate of the problem and both coefficients A_c and orbitals $\phi_{c,n}$ being time dependent. Such equation is propagated according to the Dirac–Frenkel variational principle

$$\left\langle \delta\psi \middle| \hat{H} - i\frac{\partial}{\partial t} \middle| \psi \right\rangle = 0 \quad (4.8)$$

$$\mathbf{W}_{\bar{i}} = \frac{1}{M_{\bar{i}}}\sum_{i=1\neq\bar{i}}^{I} M_i \mathbf{W}_i$$

and then $I - 1$ vectors $\mathbf{W}_i$ are sufficient to define the system.

with all orbitals being constrained to be orthonormal during propagation. At each point of the calculation, the norm and the energy of the wavepacket are monitored in order to ensure conservation. Moreover, the error introduced (into both orbitals and coefficients) by time discretization is estimated and the time step is modified accordingly. There is no restriction imposed (in principle) on the number of coordinates or their nature. Thus, one can use appropriate coordinates for the problem considered. In the following, by having chosen a spherical cavity, we can exploit the advantage of using spherical polar coordinates to describe both the electron (r, θ, ϕ) and the proton (R, Θ, Φ). Within this choice, the kinetic energy operators for the electron $\hat{T}_{el}$ and for the proton $\hat{T}_{pr}$ become separable

$$\hat{T}_{el} = -\frac{\hbar^2}{2m_e} r^{-1} \frac{\partial^2}{\partial r^2} + \frac{\Lambda_{el}}{2m_e r^2} \tag{4.9}$$

$$\hat{T}_{pr} = -\frac{\hbar^2}{2m_p} R^{-1} \frac{\partial^2}{\partial R^2} + \frac{\Lambda_{pr}}{2m_p R^2}, \tag{4.10}$$

where m_e and m_p denote the masses of the electron and of the proton respectively, r and R the related distances from the center of the sphere, and Λ_{el} and Λ_{pr} the ordinary particle-on-a-sphere angular momentum operators. One should note that the potential energy

$$V(R, \Theta, \Phi, r, \theta, \phi) = -\frac{1}{(R^2 + r^2 - 2RrC)^{1/2}} \tag{4.11}$$

with

$$C = \cos\Theta \cos\theta + \sin\Theta \sin\theta \cos(\Phi - \phi) \tag{4.12}$$

is clearly non-separable and needs to be decomposed in a suitable "sum of products". This can be done by diagonalizing an appropriate "potential density matrix" and keeping only those "natural potentials" whose populations remain above a pre-specified threshold and then integrate in time over them [45].

At this point, one can calculate the autocorrelation function of the system by evaluating the overlap integral of the wavepacket at time t with that at time t=0

$$a(t) =< \psi(0)|\psi(t) > \tag{4.13}$$

whose Fourier transform yields for the confined system a stick spectrum of its energy levels.

4.1.3 The Born–Oppenheimer Approximation

Most often, a further reduction of the complexity of Eq. (4.1) is obtained by introducing the so-called Born–Oppenheimer approximation that decouples the motion of the electrons from the motion of the nuclei (which is justified by the fact that the motion of electrons is much faster than that of the nuclei which allows them to redistribute almost instantaneously around the nuclei in motion). The Born–Oppenheimer approximation is obtained by factoring the total wavefunction $\Psi(\mathbf{W},\mathbf{w},t)$ as a product of an electronic wavefunction $\Phi(\mathbf{w};\mathbf{W})$ and a nuclear function $\Xi(\mathbf{W},t)$

$$\Psi(\mathbf{w},\mathbf{W},t) = \Phi(\mathbf{w};\mathbf{W})\Xi(\mathbf{W},t). \tag{4.14}$$

As one can see (and as physically justified above), the electronic wavefunction depends parametrically on the nuclear configuration $\mathbf{W}$ and is a solution of the Schrödinger equation for the movement of the electrons assuming fixed nuclei

$$[\mathcal{T}_e(\mathbf{W}) + \mathcal{V}_{ne}(\mathbf{w},\mathbf{W}) + \mathcal{V}_{ee}(\mathbf{w})]\,\Phi(\mathbf{w},\mathbf{W})) = E_e(\mathbf{W})\Phi(\mathbf{w},\mathbf{W}). \tag{4.15}$$

The methods used for solving this equation and determining the corresponding electronic energies, E_e, were discussed previously.

The relevant fact, for the purpose of separating nuclear motion from that of the electrons, results in the following differential equation for the motion of the nuclei (see Eq. (4.1))

$$\left[E_i(\mathbf{W}) + \hat{T}_n\right]\Xi_i(\mathbf{W},t) = i\hbar\frac{\partial}{\partial t}\Xi_i(\mathbf{W},t), \tag{4.16}$$

where E_i is the eigenvalue of the i-th potential energy surface of Eq. (4.15) in which we neglected the adiabatic corrections and the coupling between different adiabatic states which are typically small (this is the Born–Oppenheimer approximation). Hereinafter, we shall also replace E_i by V (in doing this we also drop the subscript) that is the potential energy surface (PES) governing the dynamics of the nuclei once we have selected a given Born–Oppenheimer surface.

$$\left[\hat{T}_n + V(\mathbf{W})\right]\Xi(\mathbf{W},t) = i\hbar\frac{\partial}{\partial t}\Xi(\mathbf{W},t). \tag{4.17}$$

Equation (4.17) is a differential equation of the first order in time and its solution has the form:

$$\Xi(\mathbf{W},t) = \hat{U}(t,t_0)\Xi(\mathbf{W},t_0), \tag{4.18}$$

where (as already commented for the two body systems) $\hat{U}(t,t_0)$ is the time evolution operator.

This equation can be integrated over time as an initial value problem defining the initial shape of the wave function (or wave packet) at time $t = t_0$ and applying the Hamiltonian operator until convergence of the solution.

For time-independent Hamiltonians (that is in the absence of external fields), the time dependence of the wavefunction can be factored out as follows:

$$\Xi(\mathbf{W}, t) = e^{-iEt/\hbar}\Xi(\mathbf{W}), \tag{4.19}$$

where E is the total energy of the system. Substituting this into the time-dependent Schrödinger equation and dividing by the functional dependence of time, one obtains the Schrödinger equation for stationary states of the system

$$\mathcal{H}_n \Xi(\mathbf{W}) = \left[\hat{T}_n + V(\mathbf{W})\right] \Xi(\mathbf{W}) = E\Xi(\mathbf{W}). \tag{4.20}$$

4.2 Three-Atom Systems

4.2.1 *Three-Body Orthogonal Coordinates*

The most important pros and cons of using internuclear coordinates for an atom–diatom system are easy to specify: we have already mentioned that they are particularly suited to formulate the interaction (see Fig. 4.1 for the N + N_2 reaction channel) though, as already pointed out, they do not cover homogeneously the space of molecular geometries due to the triangular rule (in other words they cannot freely vary individually) and they are not orthogonal (and are therefore less suited for quantum dynamical calculations because leading to matrices full of nonzero off-diagonal elements).

On the contrary, orthogonal coordinates have by definition the advantage of providing diagonal representations (no crossed terms) of dynamical problems. The most popular sets of orthogonal coordinates are the Jacobi ones sketched in Fig. 4.2 for the three different arrangements $\tau = 1$ (A,BC), $\tau = 2$ (B,CA), $\tau = 3$ (C,BA). More in general, the definition of the Jacobi coordinates in terms of the CM position vectors $\mathbf{W}$ is $\mathbf{R}_\tau = \mathbf{R}_{\tau,(\tau+1)(\tau+2)} = \mathbf{W}_\tau - (m_{\tau+1}\mathbf{W}_{\tau+1} + m_{\tau+2}\mathbf{W}_{\tau+2})/(m_{\tau+1} + m_{\tau+2})$ and $\mathbf{r}_\tau = \mathbf{r}_{(\tau+1)(\tau+2)} = \mathbf{W}_{\tau+1} - \mathbf{W}_{\tau+2}$ (see Fig. 4.2) with τ labeling also the isolated atoms in a modulus 3 sequence. Accordingly, the angle Θ_τ is defined as $\frac{1}{2}\arctan(\mathbf{R}_\tau\mathbf{r}_\tau)/|R_\tau - r_\tau|$.

A key feature of the Jacobi coordinates is that they are arrangement dependent (arrangements have been labeled above with different values of τ) and are therefore unsuitable for a full description of reactive processes involving a breaking of an existing bond and the forming of a new one (see Fig. 4.2).

It is often useful to scale Jacobi coordinates $\mathbf{R}_\tau$ and $\mathbf{r}_\tau$ by the mass coefficient appearing in the formulation of the kinetic operator (see some examples later in this chapter) $\mathbf{S}_\tau$ and $\mathbf{s}_\tau$ defined as $\mathbf{S}_\tau = d_\tau\mathbf{R}_\tau$ and $\mathbf{s}_\tau = d_\tau^{-1}\mathbf{r}_\tau$ in which the dimensionless scaling factor d_τ is chosen so as to stretch/compress the coordinates to the end of making more democratic their weight in the Hamiltonian (very useful for an intuitive description of isotopic effects in the graphical representation of the scattering).

The different sets of Jacobi coordinates are related by the so-called kinematic rotations (which are not physical rotations but matrix transformations relating different arrangements) like the following ones:

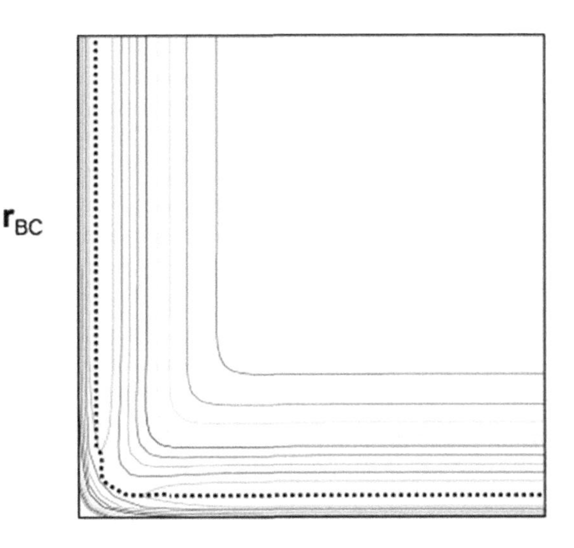

Fig. 4.1 Isoenergetic contours of the collinear N + N_2 system plotted as a function of internuclear distances evidencing the MEP (dotted line). The contours are spaced in energy of 1 eV taken from the bottom of the N_2 asymptote and show a barrier to reaction of about 1.5 eV

$$\begin{pmatrix} \mathbf{S}_{\tau+2} \\ \mathbf{s}_{\tau+2} \end{pmatrix} = \begin{pmatrix} \cos\beta_{\tau+2} - \sin\beta_{\tau+2} \\ \sin\beta_{\tau+2} + \cos\beta_{\tau+2} \end{pmatrix} \begin{pmatrix} \mathbf{S}_{\tau} \\ \mathbf{s}_{\tau} \end{pmatrix} \tag{4.21}$$

with

$$\cos\beta_{\tau+2} = \frac{m_\tau m_{\tau+2}}{(m_\tau + m_{\tau+1})(m_{\tau+1} + m_{\tau+2})} \tag{4.22}$$

and

$$\sin\beta_{\tau+2} = \frac{m_{\tau+1} M_{tot}}{(m_\tau + m_{\tau+1})(m_{\tau+1} + m_{\tau+2})} \tag{4.23}$$

with $M_{tot} = m_\tau + m_{\tau+1} + m_{\tau+2}$. Jacobi coordinates are also suitable for describing the long range interaction in which reference is made to a diatomic equilibrium geometry of the reactant and accounts for the effect of the orientation of the diatom on the polarizability of the system during the collision process.

In the early days of reactive scattering studies, the so-called natural coordinates (NC) relying on a variable (the minimum energy path (MEP) of the PES) smoothly connecting reactants and products potential energy asymptotic diatoms were proposed. NC coordinates are seemingly ideal for describing chemical reactions because they associate the MEP with two perpendicular coordinates smoothly reorienting themselves while progressing from the reactant to the product arrangement.

An example of the N + N_2 MEP (dotted line) to which the natural coordinates are defined as orthogonal at each of its points is given in Fig. 4.1 for the collinear PES of the N + N_2 (that will be examined in more detail later). The PES contours are plotted with a spacing in energy of 1 eV (taken from the bottom of the N_2 asymptote) and show a barrier to reaction of about 1.5 eV (more precisely 36 kcal/mol). The practical use of NC coordinates, however, is so cumbersome and inefficient (especially in the regions of branching between different arrangement channels) that they have never been used systematically in reactive scattering investigations.

Obviously, in order to formulate and represent the interaction, one does not need to retain the six dimensions of the two $\mathbf{R}_\tau$ and $\mathbf{r}_\tau$ vectors in the adopted functional form of the PES. To this end one can write, in fact, the equations in terms of the rotation angles (α, β and γ the three Euler angles) of the rigid three-body system (i.e., molecular plane as it has been done for transforming from center-of-mass (CM) to body-fixed (BF) formalism the two-body problem scattering equations) and formulate the PES only in terms of the other remaining three (internal) coordinates.

An alternative set of orthogonal coordinates more suitable by design to describe reactive processes is the so-called hyperspherical coordinates. Hyperspherical coordinates change the perspective of looking at a chemical process by focusing on the aggregated arrangement of the system and considering at its fragmentation into all possible products (one of which in the traditional approach of the Jacobi coordinates is considered the reactant arrangement). A particular set of hyperspherical coordinates, called Delves, at a fixed value of the Jacobi angle Θ_τ (for example collinear) consists of a hyperradius ρ that is arrangement independent and is defined as

$$\rho^2 = S_\tau^2 + s_\tau^2 \tag{4.24}$$

for any value of τ and an arrangement (τ) dependent angle θ_τ defined as

$$\theta_\tau = \arctan \frac{s_\tau}{S_\tau}. \tag{4.25}$$

In the following, we give a short list of some popular atom–diatom quantum codes made available for distribution by the authors.

ABC [46] is a time-independent atom–diatom quantum reactive scattering program using a coupled-channel hyperspherical coordinate method to solve the Schrodinger equation for the motion of the three nuclei (A, B, and C) on a single Born–Oppenheimer potential energy surface.

RWAVEPR [47] is a time-dependent atom–diatom quantum reactive scattering program using Jacobi coordinates to integrate rigorously the three-dimensional time-dependent Schrodinger equation by propagating wave packets.

DIFFREALWAVE [48] is a parallel real wavepacket code for the quantum mechanical calculation of reactive state-to-state differential cross sections in atom–diatom collisions using Jacobi coordinates and a real wavepacket.

A more general set of Adiabatically adjusting coordinates which follow the evolution of the Principal axis of inertia of the system Hyperspherical coordinates (APH) [49] that will be discussed later has also been proposed and the related codes have also been made available for circulation.

Another set of programs, of which some versions have been extended to more than three atoms, is FLUSS-MCTDH [45, 51]. This is a pair of programs carrying out a multiconfiguration time-dependent Hartree (MCTDH) calculation of thermally averaged quantum dynamics properties of multidimensional systems based on a modified Lanczos iterative diagonalization of the thermal flux operator.

4.2.2 Atom–Diatom Reactive Scattering Jacobi Method

We have discussed earlier about the suitability of any kind of coordinates for classical mechanics treatments. As already mentioned, for time-independent quantum treatments, Jacobi coordinates, which are defined starting from the vectors connecting the centers-of-mass of the fragments of the system considered (see Fig. 4.2 for the particular case of the system atom–diatom), are well suited only for nonreactive processes because initial and final molecular fragments coincide and therefore R is a good continuity variable.

For reactive processes, instead, Jacobi coordinates lead to some difficulties when switching from reagent to product formulations.

For arrangement conserving elementary processes (the nonreactive ones), the quantities of experimental interest that can be usually associated with theoretical treatments are

- the population of the final states of the products measured by a spectrometer (that are amenable to the process probability),
- the intensity of matter collected at a certain solid angles by the detector in crossed molecular beam apparatus (that is amenable to the process cross section),

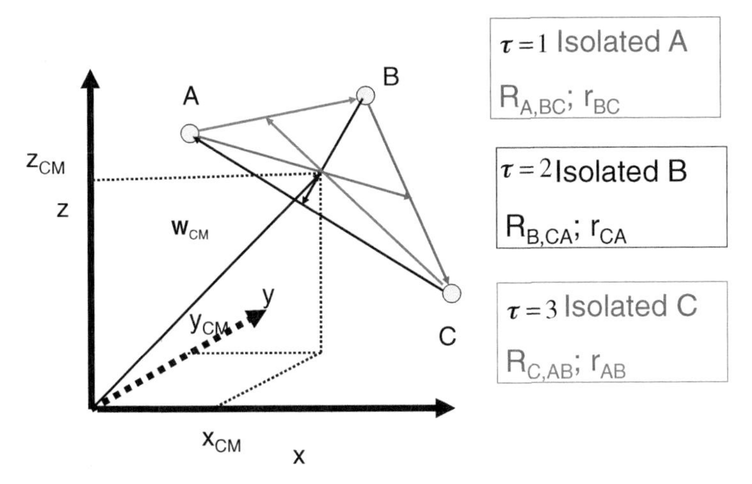

Fig. 4.2 Jacobi coordinates defined for different arrangements of the particles A, B, and C

- the measurement of the velocity of a reaction (that is amenable to the integration of the cross section over the energy thermal distribution).

Currently, exact quantum calculations have been implemented mainly for collisions of three atoms (and seldom for systems with four or a higher number of atoms). In the particular case of a three-atom system, one has (in atomic units and eliminating the label τ for the sake of simplicity)

$$\hat{H}\Phi^{J\lambda}(R,r,\Theta,t) = \left\{-\frac{1}{2\mu_R}\frac{\partial^2}{\partial R^2} - \frac{1}{2\mu_r}\frac{\partial^2}{\partial r^2}\right\}\Phi^{J\lambda}(R,r,\Theta,t) - \left(\frac{1}{2\mu_R R^2} + \frac{1}{2\mu_r r^2}\right)\left\{\frac{1}{\sin\Theta}\frac{\partial}{\partial\Theta}\sin\Theta\frac{\partial}{\partial\Theta} - \frac{\lambda^2}{\sin^2\Theta}\right\}\Phi^{J\lambda}(R,r,\Theta,t) + \frac{1}{2\mu_R R^2}\{J(J+1) - 2\lambda^2\}\Phi^{J\lambda}(R,r,\Theta,t) + V(R,r,\Theta)\Phi^{J\lambda}(R,r,\Theta,t) + C^J_{\lambda,\lambda-1}\Phi^{J,\lambda-1}(R,r,\Theta,t) + C^J_{\lambda,\lambda+1}\Phi^{J,\lambda+1}(R,r,\Theta,t), \tag{4.26}$$

where R, r, and Θ are, indeed, the Jacobi coordinates. In Eq. (4.26), the terms

$$C^J_{\lambda,\lambda\pm1} = -\frac{[J(J+1) - \lambda(\lambda\pm1)]^{\frac{1}{2}}[j(j+1) - \lambda(\lambda\pm1)]^{\frac{1}{2}}}{R^2} \tag{4.27}$$

are respectively raising and lowering operators of the total angular momentum $\mathbf{J}$ (whose quantum number is J) and when $J = 0$ the last three C^J terms are zero [52]. The above set of equations refer to the Hamiltonian formulated in a BF coordinate system where the z-axis is pointing toward the atom. In that case, λ is the projection of the total angular momentum on the body-fixed z_λ axis. The additional coupling terms $C^J_{\lambda,\lambda\pm1}$ appear because we are using a body-fixed coordinate system.

To go back to the problem of the inadequacy of using Jacobi coordinates for time-independent quantum reactive scattering calculations, we emphasize here the fact that this is less a problem when using TD techniques. In fact, related time-dependent methods can be used provided that, once generated the system wavepacket of the reactants in the related coordinates, it can be converted into the product ones and the equations can be integrated using the Fourier transform method for the radial part (R and r) and the discrete variable representation (DVR) for the angular coordinate [53]. For this purpose, the initial wave function is evaluated on the grid points and the repetitive application of the time evolution operator results in a time-dependent snapshot of the evolving wavepacket [52].

The time-dependent coefficients of the basis functions have the form

$$C_{vj\lambda,v'j'\lambda'}(t) = \int_{r'} \mathrm{d}r' \int_{\theta'} \mathrm{d}\Theta' \sin\Theta' P_{r'\lambda'}(\theta')\phi_{v'j'}(r')\Psi^{J\lambda'}(R = R_\infty, r, \theta, t) \tag{4.28}$$

(primed coordinates are those of the products while, the unprimed ones are those of the reagents). In Eq. (4.28), $P_{r'\lambda'}(\theta')$ is the associated Legendre polynomial the angular part of the wave function of the rotational state j' of the products. From the Fourier transform of the coefficients C, one can obtain the energy dependent coefficients A

$$A_{\nu j\lambda,\nu' j'\lambda'}(E) = \frac{1}{2\pi}\int_{t=0}^{\infty} \mathrm{d}t\; exp\,(iEt/\hbar)\cdot C_{\nu j\lambda,\nu' j'\lambda'}(t) \tag{4.29}$$

from which the reactive scattering **S** matrix element is numerically evaluated using the relationship

$$S_{\nu j\lambda,\nu' j'\lambda'}(E) = \left(\frac{k_{\nu j}}{\mu\mu'}\right)^{1/2} \frac{\hbar}{g(-k_{\nu j})} e^{-k_{\nu' j'}R_{\infty}} A_{\nu j\lambda,\nu' j'\lambda'}(E). \tag{4.30}$$

4.2.3 *Atom–Diatom Time-Independent APH Method*

The above-mentioned transformation from the reactant to the product formalism can be avoided using the so-called hyperspherical coordinates. In order to illustrate the hyperspherical coordinates, let us consider first the collinear (three atoms on a row) case and, in particular, the $H + H_2$ system.

Figure 4.3 shows the isoenergetic contours of the $H + H_2$ PES system as a function of the Jacobi R_α and r_α coordinates. In the same figure, the related collinear hyperspherical coordinates $\rho = \sqrt{R^2 + r^2}$ and the arrangement channel label $\alpha = \arctan r_\alpha/R_\alpha$ are shown while the corresponding fixed ρ cuts are plotted

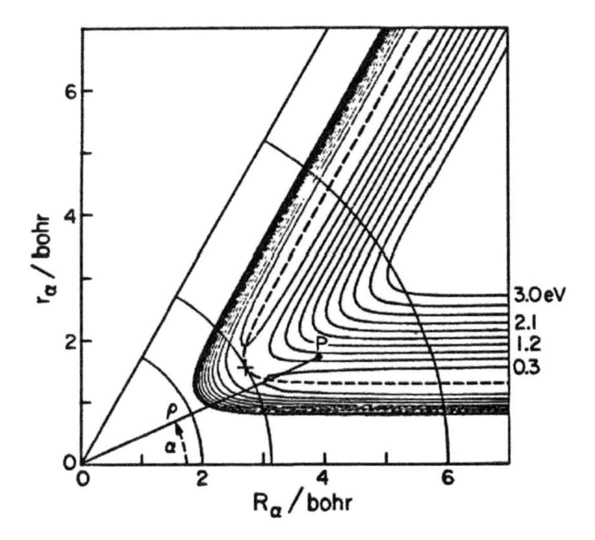

Fig. 4.3 Isoenergetic contours of the $H + H_2$ system PES and the related Jacobi and hyperspherical coordinates

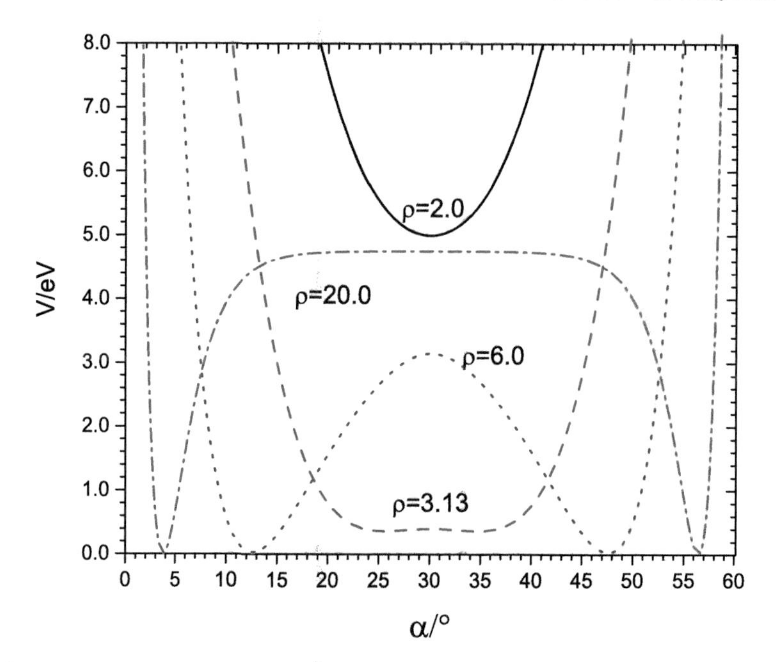

Fig. 4.4 Fixed ρ (in a.u.) cuts of the collinear $H + H_2$ PES

in Fig. 4.4 starting from a value associated with the high repulsive wall and moving then to the ρ value associated to the saddle of the reactive process and next to asymptotic like regions. In Fig. 4.3, it is apparent that the skewing angle effect that is associated with the light nature of the exchanged mass (though, being the three masses equal, the effect is less pronounced than in the case of a light mass exchanged between two heavy masses (like the exchange of H between two T isotopes or halogens)). However, the skewing angle effect shown by Fig. 4.3 is much larger than that shown by the system $H + Cl_2$ that we shall consider later for which the heavy atom Cl is exchanged between the light H atom (of HCl) and the heavy Cl atom of Cl_2.

In Fig. 4.5, the fixed $\rho = 6$ a.u. eigenfunctions (used to describe the bound motion of H between the other two H atoms in $H + H_2$) are shown. As apparent from the plot, the small ρ cuts have the single well shape while the separate channel structure of large ρ values is characterized by double well shapes. On these cuts are calculated the fixed ρ surface functions whose eigenvalues give rise to the adiabats along which the reactive flux takes place.

In order to cope with the more complex nature of reactive processes starting from reactants and branching into different product channels, one can use the already mentioned 3D Adiabatically adjusting Principal axis of inertia Hyperspherical APH coordinates [49] whose hyperradius is a reaction coordinate that has the advantage of unifying reactive and nonreactive processes (ρ is the radius of the hypersphere that is subtended by all Jacobi coordinates).

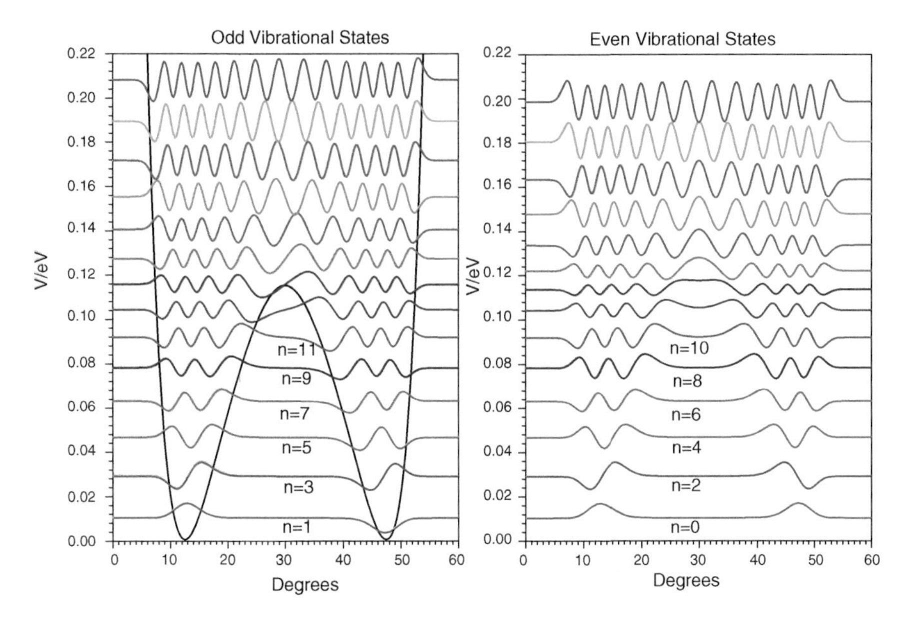

Fig. 4.5 Eigenvalues of the hyperangular bound states of a fixed large ρ cut of the collinear H + H_2 PES

At a constant value of the hyperradius, the functions of the other variables (all of angular type) can be calculated by solving a problem eigenvalue which has a dimensionality to one less than the overall problem. These eigenfunctions of the hyperangles are then used as the basis for the development of the global function in a coupled channel (CC) approach. This allows you to split the computational procedure in a first step which consists in the calculation of the mentioned hyperangular eigenfunctions and in a second step in the propagation of the solution to values near ρ=0 (all particles collapse) to large values (fragmentation into different forms depending on the values of the hyperangles). The final step is devoted to the comparison of the propagated function with its asymptotic form and then to the calculation of the scattering **S** matrix.

In the APH approach [49], the two internal angles θ and χ are defined as

$$\tan\theta = \frac{\left[(S_\tau^2 - s_\tau^2)^2 + (2\vec{S}_\tau \cdot \vec{s}_\tau)^2\right]^{\frac{1}{2}}}{2S_\tau s_\tau \sin\Theta_\tau} = \frac{\left[\cos^2 2\theta_d + \sin^2 2\theta_d \cos^2\Theta\right]^{\frac{1}{2}}}{\sin 2\theta_d \sin\Theta} \tag{4.31}$$

and

$$\sin 2\chi_\tau = \frac{2\vec{S}_\tau \cdot \vec{s}_\tau}{\left[(S_\tau^2 - s_\tau^2)^2 + (2\vec{S}_\tau \cdot \vec{s}_\tau)^2\right]^{1/2}} = \frac{\sin 2\theta_d \cos\Theta}{\left[\cos^2 2\theta_d + \sin^2 2\theta_d \cos^2\Theta\right]^{\frac{1}{2}}} \tag{4.32}$$

$$\cos 2\chi_{\tau} = \frac{S_{\tau}^2 - s_{\tau}^2}{\left[(S_{\tau}^2 - s_{\tau}^2)^2 + (2\vec{S}_{\tau} \cdot \vec{s}_{\tau})^2\right]^{1/2}} = \frac{\cos 2\theta_d}{\left[\cos^2 2\theta_d + \sin^2 2\theta_d \cos^2 \Theta\right]^{\frac{1}{2}}} \tag{4.33}$$

and the three Euler angles $(\alpha_Q, \beta_Q, \gamma_Q)^2$ define the location of the colliding bodies on the fixed $\rho = \sqrt{S_{\tau}^2 + s_{\tau}^2}$ hypersurface.

It should be noted, here, that ρ and θ are truly independent of τ contrary to the Delves θ_{τ}. Since the APH coordinates treat all arrangement channels democratically it makes no difference which initial τ we use. In fact, to change from an arrangement channel labeled by τ to one labeled by $\tau + 1$, one simply rotates the angle χ_{τ}, i.e., $\chi_{\tau+1} = \chi_{\tau} + \chi_{\tau,\tau+1}$. The angle $\chi_{\tau,\tau+1}$ is a constant that only depends on the masses of the three particles and the channel to channel rotation angles are

$$\cos \chi_{\tau+1,\tau} = -\frac{\mu}{d_{\tau} d_{\tau+1} m_{\tau+2}} \quad \text{and} \quad \sin \chi_{\tau+1,\tau} = -\frac{1}{d_{\tau} d_{\tau+1}}, \tag{4.34}$$

where the constants

$$d_{\tau} = \left[\frac{m_{\tau}}{\mu}\left(1 - \frac{m_{\tau}}{M_{tot}}\right)\right] \tag{4.35}$$

with the three particle reduced mass $\mu = \sqrt{m_A m_B m_C / M_{tot}}$ and total mass $M_{tot} = m_A + m_B + m_C$. The dependence of χ by the choice of the reference geometry τ will be neglected hereafter for simplicity omitting the corresponding subscript.

The hyperradius ρ determines the overall size of the three particle system, θ is a bending angle, and χ is a kinematic rotation angle. Using these coordinates, the equations for the internal coordinates have the form

$$\begin{aligned} \left[T_{\rho} + T_h + T_r + T_c + V\right] \; \Psi^{JMp}(\rho, \theta, \chi, \alpha_Q, \beta_Q, \gamma_Q) \\ = E\Psi^{JMp}(\rho, \theta, \chi, \alpha_Q, \beta_Q, \gamma_Q) \,, \end{aligned} \tag{4.36}$$

where p is the parity of the system, M is the projection of the total angular momentum on the space fixed z-axis. The three physical properties J, M, and p are conserved quantities in the Schrödinger equation. In the equation, "h", "r", and "c" are respectively for "hypersphere", "rotational", and "Coriolis" and the symbols T_{ρ}, T_h, T_c, and T_r have the form

$$T_{\rho} = -\frac{\hbar^2}{2\mu\rho^5}\frac{\partial}{\partial\rho}\rho^5\frac{\partial}{\partial\rho} = -\frac{\hbar^2}{2\rho^{5/2}}\frac{d^2}{d\rho^2}\rho^{5/2} + \frac{15}{8\mu\rho^2},$$

$$T_h = -\frac{\hbar^2}{2\mu\rho^2}\left(\frac{4}{\sin 2\theta}\frac{\partial}{\partial\theta}\sin 2\theta\frac{\partial}{\partial\theta} + \frac{1}{\sin^2\theta}\frac{\partial^2}{\partial\chi^2}\right)$$

[2]Which rotate the coordinates (passive rotations) to body-fixed coordinates where the z_Q axis points along the smallest principle moment of inertia and the y_Q is perpendicular to the plane formed by the 3-particle system. For simplicity, we will hereafter drop the subscript Q.

$$T_r = A(\rho,\theta)J_x^2 + B(\rho,\theta)J_y^2 + C(\rho,\theta)J_z^2, \tag{4.37}$$

$$= \frac{A+B}{2}J^2 + \frac{A-B}{2}(J_x^2 - J_y^2) + \left[C - \frac{A+B}{2}\right]J_z^2 \tag{4.38}$$

and

$$T_c = -\frac{i\hbar\cos\theta}{\mu\rho^2\sin^2\theta}J_y\frac{\partial}{\partial\chi},$$

where inverses of $A(\rho,\theta)$, $B(\rho,\theta)$, and $C(\rho,\theta)$ are defined as $A^{-1}(\rho,\theta) = \mu\rho^2(1 + \sin\theta)$, $B^{-1}(\rho,\theta) = 2\mu\rho^2\sin^2\theta$, $C^{-1}(\rho,\theta) = \mu\rho^2(1 - \sin\theta)$.

Since there are no external fields, the interaction potential is independent of its orientation in space and thus independent of the three Euler angles, i.e., $V = V(\rho,\theta,\chi)$. Now we need some basis functions. We define our basis functions as a product of analytic Wigner rotation functions times a numerically calculated surface function (called surface function because it is the solution or wave function on the surface of a fixed hypersphere of radius ρ) and use a linear combination of these surface functions to expand the wave function in each sector i. The surface functions at each ρ_i (i.e., the value of ρ at the midpoint of sector i), $\Phi_{t\Lambda}^{Jp}$, are the solutions of the following Hamiltonian:

$$\left[T_h + \frac{15\hbar^2}{8\mu\rho_i^2} + C(\rho_i,\theta)\hbar^2\Lambda^2 + V(\rho_i,\theta,\chi) - \varepsilon_{t\Lambda}^{Jp}(\rho_i)\right]\Phi_{t\Lambda}^{Jp}(\theta,\chi;\rho_i) = 0 \tag{4.39}$$

where t is an index to label the t-th eigenenergy and Λ is the projection of the total angular momentum on the z component of the body-fixed axis (note that the $\frac{15}{8\mu\rho^2}$ kinetic energy term is included in the surface function Hamiltonian) (Fig. 4.6).

Once all of the surface functions have been calculated and related eigenvalues have been determined for all the ρ_i values considered (related plots as a function of ρ are called adiabats because connecting adiabatically such eigenvalues), we obtain the following set of coupled second-order differential equations in ρ

$$\left[\frac{\partial^2}{\partial\rho^2} + k^2\right]\psi_{t\Lambda}^{Jpn}(\rho) =$$

$$\frac{2\mu}{\hbar^2}\sum_{t'\Lambda'} < \Phi_{t\Lambda}^{Jp}(\theta,\chi,\rho_i)D_{\Lambda M}^{\hat{J}p}|H_{int}|\Phi_{t\Lambda'}^{Jp}(\theta,\chi,\rho_i)D_{\Lambda M}^{\hat{J}p} > \psi_{t\Lambda'}^{Jpn}(\rho), \tag{4.40}$$

where as usual $k^2 = 2\mu E/\hbar^2$. In Eq. (4.40), the internal Hamiltonian has the form H_{int}

$$H_{int} = T_h + T_c + \frac{15\hbar^2}{8\mu\rho^2} + V(\rho,\theta,\chi) \tag{4.41}$$

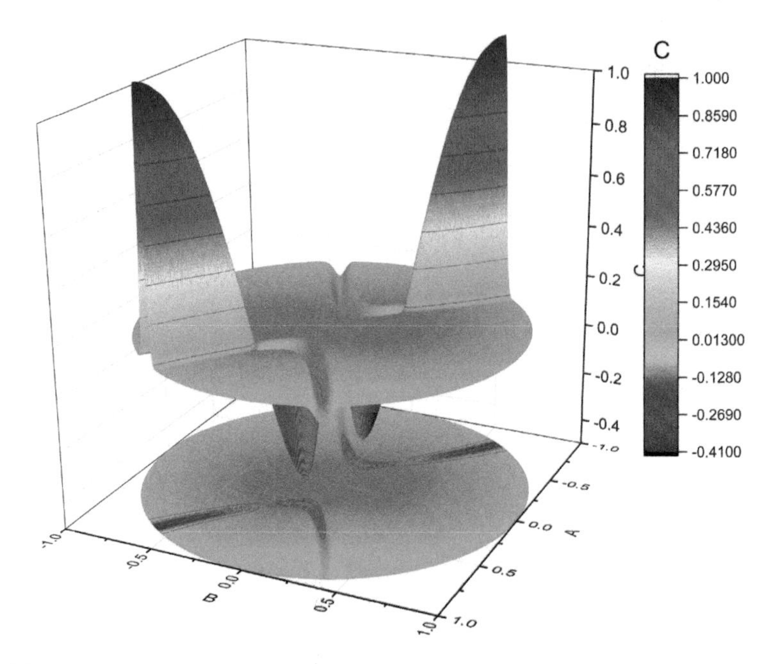

Fig. 4.6 A typical plot of a (fixed ρ 12.0 bohr) surface function Φ located in the product channel (LiF) of the reaction Li + HF $\rightarrow$ LiF + H taken at $v = 0$ (no nodes for the LiF vibration along the arc) and $j = 1$ (there is one node for the LiF rotation along the radius). Isometric contours of the surface function are projected on the plane below

As a consequence, the relative computational procedure is often divided into three parts. Of these, the first part is devoted to the calculation of the surface functions $\Phi^{Jp}{}_{t\Lambda}(\theta, \chi; \rho)$ which depend on θ and χ and parametrically on the ρ_i grid. For three atoms, the two-dimensional Hamiltonian and surface functions only depend on two angular coordinates (θ, χ).

The second part of the calculation is to propagate the logarithmic derivative[3] (which is very useful in keeping the solution numerically stable.) of the radial wave function using a matrix of coupled second-order differential equations from the origin where the wave function is zero to a large hyperradius where asymptotic boundary conditions can be applied. This part of the procedure is dominated by the inversion of matrices of size equal to the number of basis functions used to expand the scattering wavefunction.

In the third part of the computation, one faces the task of projecting the computed solution onto the Jacobi asymptotic region in order to extract the scattering matrix **S**

[3]The logarithmic derivative is

$$y = \frac{d\psi}{d\rho}\psi^{-1}. \tag{4.42}$$

matrix. Currently, there exist implementations of this method to treat the reactions of quantum systems of either three and four atoms.

The corresponding code for triatomic systems, APH3D [49] is available for distribution.

4.2.4 The Atom–Diatom Time-Dependent APH Method

Time-independent methodologies tend to deal with characteristic properties of the system under study like ground state equilibrium properties, energy level structure, and statistical thermodynamical aspects which are invariant to initial conditions set by the experiment. On the contrary, one may want to study the properties of a system prepared in specific initial conditions. In this case, time-dependent calculations are typically the way to proceed.

The formalism of the APH coordinates can also be used for a time-dependent approach. In this case, we start with the Hamiltonian of Eq. (4.36) and instead of calculating surface functions that parametrically depended on ρ one discretizes the hyperradius, and the two internal angles θ and χ. To this end, one can use an equally spaced grid in ρ, a DVR grid in θ, a periodic DAF grid in χ and the analytic Wigner rotation functions for the three Euler angles. In this way, one ends up with the very large set of coupled time-dependent Schrödinger equations

$$- i\hbar \frac{\partial \psi^{J\Lambda}(\rho, \theta, \chi, t)}{\partial t} = \mathbf{H}\psi^{J\Lambda}(\rho, \theta, \chi, t) \tag{4.43}$$

of dimension $N_\rho N_{DVR} N_\chi (J + 1)$. For time-independent Hamitonians, we use the time evolution operator

$$\psi^{J\Lambda}(\rho, \theta, \chi, t) = \mathbf{U}(t, t_0)\psi^{J\Lambda}(\rho, \theta, \chi, t) = e^{-i\frac{\mathbf{H}t}{\hbar}}\psi^{J\Lambda}(\rho, \theta, \chi, t_0), \tag{4.44}$$

where $\psi^{J\Lambda}(\rho, \theta, \chi, t_0)$ is any desired initial wavepacket (that is often taken as a linear combination of initial quantum states with an initial Gaussian energy distribution and momentum) and the exponential of the $\mathbf{H}$ matrix is evaluated using the Taylor series expansion of the exponential

$$\exp[x] = 1 + x + \frac{x^2}{2!} + \frac{x^3}{3!} + \frac{x^4}{4!} + \frac{x^5}{5!} + \cdots \tag{4.45}$$

where in our case

$$x = -i\frac{\mathbf{H}}{\hbar}t \tag{4.46}$$

In actuality, one can use a Chebychev economization of the Taylor series expansion to decrease the number of terms necessary for convergence. As the wavepacket propagates through a large hypersphere ρ_{max}, the wavepacket is projected onto surface functions evaluated at ρ_{max}. This provides a time-dependent snapshot of these projections. Finally, these snapshots are then Fourier transformed to extract the energy-dependent **S** matrix elements (as already discussed for the Jacobi time-dependent method).

The corresponding code, TD-APH3D [54] is available for distribution.

4.3 Beyond Full Quantum Calculations

4.3.1 Reduced Dimensionality Quantum Treatments

For a variety of reasons, you may not want or can undertake a full quantum dynamical study of the investigated process. Because of this, you might be ready to carry out the calculations by introducing some approximations (like adopting a reduced dimensionality treatment). This is especially convenient when one needs to lower computational costs, is ready to accept a lower level of accuracy, or has in any case to highly average detailed outcomes (like estimate a thermally average rate coefficient).

A first way of reducing the dimensionality of the accurate treatment of an atom–diatom reaction is to limit its use to the lowest value of the quantum numbers (e.g., perform only the zero total angular momentum J) calculations and then extrapolate or model the behavior for higher values. In the case of J, one can model the higher total angular momentum results by assuming that an increase of the total angular momentum of the system plays the role of freezing the corresponding quantity of energy and simply shifting up the state-specific probability along the energy scale (J shifting). More formally, one can write

$$P_i^{J,\Lambda} \simeq P_i^{J=0}(E - E_{J,\Lambda,i}^{\pm}) \tag{4.47}$$

in which for a symmetric top $E_{J,\Lambda,i}^{\pm}$ is given by

$$E_{J,\Lambda,i}^{\pm} = \bar{B}_i^{\pm} J(J+1) + (A_i^{\pm} - \bar{B}_i^{\pm})\Lambda^2 \tag{4.48}$$

Other dimensionality reductions can be introduced by decoupling some degrees of freedom. Two main families of reduced dimensionality treatments have been developed: sudden [55, 56] and adiabatic [57] ones and have been sometimes extended and adapted to larger systems (in particular four-atom ones).

In the sudden scheme, conditions are considered to change rapidly. Accordingly, the system is prevented from adapting its configuration during the process, hence the spatial probability density remains unchanged. Typically, there is no eigenstate of the final Hamiltonian with the same functional form as the initial state. The system ends in a linear combination of states that sum to reproduce the initial probability density.

In the adiabatic scheme, conditions are considered to change gradually. Accordingly, the system is allowed to adapt its configuration during the process, hence the probability density is modified by the process. Typically, the system starts in an eigenstate of the initial Hamiltonian, it will end in the corresponding eigenstate of the final Hamiltonian

Of the first type is the popular infinite order sudden approximation (IOSA) that applies, at the same time, an energy and a centrifugal sudden dimensionality reduction to the full three-dimensional Eq. (4.26) formulation of the scattering equations. The IOSA scheme in Jacobi coordinates leads to the the following set of fixed collision angle Θ_τ and fixed reactants' orbital angular momentum l_τ two-dimensional equations:

$$\left[-\frac{\hbar^2}{2\mu}\left(\frac{\partial^2}{\partial S_\tau^2}+\frac{\partial^2}{\partial s_\tau^2}-T_l-T_r\right)+V(S_\tau,s_\tau;\Theta_\tau)-E\right]\Phi_\tau^{l_\tau}(S_\tau,s_\tau;\Theta_\tau)=0.$$

In Eq. (4.49), the mass scaled Jacobi coordinates $S_\tau=(\mu_\tau/\mu)^{1/2}R_\tau$ and $s_\tau=(m_\tau/\mu)^{1/2}r_\tau$ of arrangement τ are used to compute the fixed $T_l=l_\tau(l_\tau+1)/S_\tau^2$ and $T_r=j_\tau(j_\tau+1)/s_\tau^2$ **S** matrix elements.

For the particular case of collinear atom–diatom collisions ($\Theta_\tau=180°$, $l_\tau=0$ and $j_\tau=0$), Eq. (4.49) takes the particularly simple form

$$\left[-\frac{\hbar^2}{2\mu}\left(\frac{\partial^2}{\partial S_\tau^2}+\frac{\partial^2}{\partial s_\tau^2}\right)-V(S_\tau,s_\tau;\Theta_\tau)-E\right]\Phi_\tau(S_\tau,s_\tau;\Theta_\tau)=0 \tag{4.49}$$

making the matching between entrance and exit channel exact and the calculation of the s_τ component of the wavefunction of both channels as simple as the solution of the one-dimensional (fixed S_τ fixed Θ_τ) eigenvalue problem

$$\left[-\frac{\hbar^2}{2\mu}\frac{\partial^2}{\partial s_\tau^2}-V(s_\tau;S_\tau,\Theta_\tau)-\varepsilon_{\tau v}\right]\phi_{\tau v}(s_\tau;S_\tau,\Theta_\tau)=0 \tag{4.50}$$

in which, as already mentioned, $\Theta_\tau=180°$ and S_τ is segmented in many small intervals (boxes) driving the solution from the strong interaction region (were polar coordinates centered on an energetically inaccessible point of the PES ridge are used) to the asymptotic regions (where Cartesian coordinates are used).

Although seemingly too far from the real molecular 3D world, as we shall see in the next subsections, the collinear case is particularly instructive for the atom–diatom reactive phenomenology. To the end of constructing the numerical solution of Eq. (4.49) $\Phi_{\tau v}$, $(S_\tau,s_\tau;\Theta_\tau)$ is expressed as a product of $\phi_{\tau v}(s_\tau;S_\tau,\Theta_\tau)$ and $\chi(S_\tau)$

and in each box i the potential is assumed to be constant along the propagation coordinates. Accordingly, the equations resulting from such expansion (and integration over s_{τ}) are

$$\left[-\frac{\hbar^2}{2\mu}\frac{d^2}{\partial S_{\tau}^2} - D_{\tau}^i\right]\phi_{\tau v}(s_{\tau};\, S_{\tau}, \Theta_{\tau}) = 0 \tag{4.51}$$

which are integrated up to the asymptotes where by imposing the scattering boundary conditions the $S_{vv'}(E)$ matrix elements are evaluated (whose square modulus is the reaction probability) and combined to compute the reactive scattering cross section.

Of the second type is the adiabatic treatment of the overall rotational energy in which following the formalism of J.M. Bowman [58] the Hamiltonian is then given by

$$H^{J,\Lambda} = H_{eff}^{J=0} + E^{J,\Lambda}(\mathbf{Q}), \tag{4.52}$$

where $E^{J,\Lambda}(\mathbf{Q})$ is the rotational energy calculated at each nuclear configuration, denoted $\mathbf{Q}$. In many cases, Λ is nearly a good quantum number and the symmetric top expression may be used, e.g., for a prolate symmetric top

$$E^{J,\Lambda}(\mathbf{Q}) = \bar{B}(\mathbf{Q})J(J+1) + [A(\mathbf{Q}) - \bar{B}(\mathbf{Q})]\Lambda^2 \tag{4.53}$$

4.3.2 *Leveraging on Classical Mechanics*

Once abandoned the idea of carrying out accurate quantum calculations, the simplest approach in terms of formalism and computer demand is the use of a classical mechanics treatment. After all, if you do not need to reproduce in full the rich structure of a quantum calculation trajectory calculations are of great help even when for numerical reasons their integration is only partially successful in terms of energy andor angular momentum conservation. This is singled out by the comparison of the $J = 0$ exact quantum $\mathbf{P}_{00}^{J=0}$ probability computed at $v, j = 0, 0$ with the corresponding one obtained from classical mechanics for the $N + N_2$ reaction given in Fig. 4.7 (this reaction will be considered again when comparing the values of the thermal rate coefficient of the $N + N_2$ reaction computed using quantum reactive IOSA, semiclassical and quasiclassical techniques, among them and with experimental data. The figure tells us that trajectory calculations are able, indeed, to reproduce the average trend (and when using moderate rejection criteria for discarding poor energy conserving trajectories, also the absolute value) of quantum reactive probabilities even in just above the threshold energy region. On the contrary, they are unable to reproduce the detailed structure of the quantum reactive probability. Therefore, while the use of quasiclassical probabilities and cross sections in a multiscale application may not be necessarily safe in some cases, the use of more averaged quantities does not

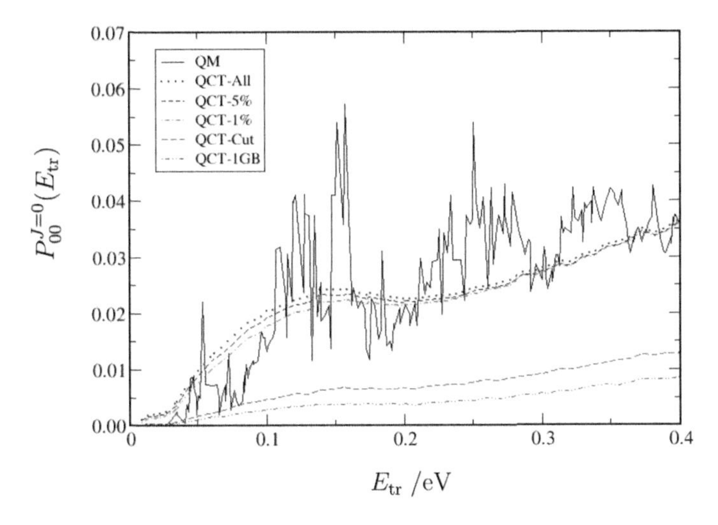

Fig. 4.7 Plot of the quantum (solid line) and quasiclassical probabilities for the N + N_2 reaction associated with trajectory calculations (various discontinuous lines as specified in the figure) adopting different criteria for discarding poorly energy conserving results

only offer the advantage of being extremely efficient on a distributed computing infrastructure but ensures also an acceptable level of accuracy.

Other accuracy constraints can be weakened when moving to more complex systems and this fuels additional interest in the adoption of trajectory techniques. This is also the case of the already considered situations in which the Born–Oppenheimer assumption breaks down and quantum calculations are too cumbersome.

In the first case, the coupling of nuclei and electronic motion is regained by allowing nonelectronically adiabatic events to occur. This is the case of strong coupling of large-amplitude molecular motion with electron degrees of freedom. In the most popular ways of dealing with these problems, nuclei are assumed to move classically on a single potential energy surface until an avoided surface crossing (in the electronically adiabatic approach) or other regions of large nonadiabatic coupling is reached. At such points, the trajectory is allowed to branch over different paths and progress on different PESs. This model treatment has been applied to different systems and its validity has been assessed by numerical integration of the appropriate semiclassical equations [59]. A large number of three-dimensional trajectory surface hopping treatments have been reported in the literature. Derivation and numerical tests of mixed quantum-classical schemes to deal with such nonadiabatic processes have also been reported and approximations to the exact coupled dynamics of electrons and nuclei offered by the factorization of the electron–nuclear wave function have been discussed [60].

The second case is the already mentioned use of SC-IVR formulation of the **S** matrix which, while preserving the ability of discretizing internal energy, leverage is still entirely based on the regaining of the whole classical trajectory information [61].

4.3.3 Semiclassical Treatments

As mentioned above and already discussed in the first chapter, there are quantities of classical nature (such as the classical action) allowing to regain quantum-like effects from the outcomes of trajectory calculations. One can, in fact, formulate the **S** matrix elements in a semiclassical, SC, fashion by working out of classical quantities the features needed to build the semiclassical wavefunction of the considered system.

At the root of the SC approach to chemical processes is the Jeffreys, Kramer, Brilluoin, and Wentzel (JWKB [62–65]) solution of the one-dimensional $l = 0$ Schrödinger equation (2.27) once it is written as

$$\left[\frac{\mathrm{d}^2}{\mathrm{d}r^2} + \frac{p^2}{\hbar^2}\right]\psi(r) = 0 \tag{4.54}$$

by assuming $p(r) = [2\mu(E - V(r))]^{1/2}$ to be real. The solution of Eq. (4.54) is $\psi(r) = Ae^{\pm ip/\hbar}$ if $p(r)$ is independent of r. This is not true, as is usually the case. In the case of $p(r)$ varying slowly with r one can make the position $\psi(r) = Ae^{\pm iS/\hbar}$ (with $S = \int p(r)dr$ being the classical action integral) whose second derivative is $\psi''(r) = \left[-(S'/\hbar)^2 \pm iS''/\hbar\right]\psi''(r)$. By expanding the JWKB wavefunction in series of $\hbar$ ($S(r) = S_o(r) + \hbar S_1(r) + \hbar^2 S_2(r)$) and equating to zero in succession its terms one gets also higher order solutions. It has to be noted here that the derivation does not place the requirement for $p(r)$ to be real. Accordingly, together with the general classically allowed solution

$$\psi(r) = A[p(r)]^{-1/2}e^{\pm iS/\hbar} \tag{4.55}$$

for $p(r)$ positive one can write also the classically forbidden one

$$\psi(r) = A[p(r)]^{-1/2}\cos\left(\frac{1}{\hbar}\int_{r_o}^{r} p(r')dr' + \alpha\right), \tag{4.56}$$

where $[p(r)]^{-1/2}$ relates the amplitude of the wavefunction (or better its square modulus) to the time spent to cross the element dr. Accordingly, the JWKB wavefunction diverges at any classical turning point.

This can be avoided by resorting to uniform approximations in which the wavefunction is imposed a shape dictated by the locations in r of the classical turning points of the potential. For example, one can assume that the wavefunction depends on a given function of r and has built in, therefore, a particular behavior in r (the wavefunction dies away in the classically forbidden region and oscillates in the classically allowed one). This allows one to handle in terms of special functions the problem of isolated turning points, potential wells, potential barriers, potential singularities, etc., and in terms of a family of quantum numbers the bound states. In particular, the Bohr–Sommerfield rule can be adopted in order to quantize the action (see Eq. (1.52)) associated with vibrations between the two turning points (say a and b) as follows

$$\Delta = \oint_a^b p(x)dx = (v + \frac{1}{2})h. \tag{4.57}$$

The area covered by the integrand of Eq. (4.57) indicates the phase space associated with the bound motion on the potential considered. In order to formulate the semiclassical **S** matrix[4] when considering all the degrees of freedom in non-separable processes (like in the case either of the inclusion of the electronic structure of the colliding partners or of the reactive or nonreactive atom oscillating diatom collision), one naturally turns, as already illustrated in the previous subsections, into the S matrix relating the family of events linking the initial state to the desired final ones (each bearing in the semiclassical approach an associated phase determined by the classical action defined in Chap. 1 accumulated along its path). For the sake of simplicity, it is assumed that the (single) translational variable (in the equation below and in Fig. 4.8) is r and that the related motion is coupled to a single internal degree of freedom α that is the angle to which is associated the action integral Δ (defined as in Eq. (4.57))

[4]The semiclassical connection between the deflection angle θ and the JWKB approximation to the phase shift δ_l can be obtained from the semiclassical formulation of the wavefunction

$$\psi_l(r) \overset{r\to\infty}{\sim} \sin\left(kr - l\pi/2 + \delta_l\right) \tag{4.58}$$

that gives

$$\delta_l \overset{r\to\infty}{\sim} \left(\int_a^r k_l(r)dr - kr + (l + 1/2)\pi/2 \right), \tag{4.59}$$

where $k_l(r)$ is the Langer-corrected wavenumber function (resulting from the transformation of r into e^x and to the scaling of the wavefunction by $e^{x/2}$ to the end of taking into account the singularity occurring at $\theta = 0$) that reads

$$k_l(r) = \left[k^2 - U(r) - \frac{(l + 1/2)^2}{r^2} \right]^{1/2} \tag{4.60}$$

from which a comparison with the quantum solution gives

$$\theta_l = 2(\partial\delta_l/\partial l). \tag{4.61}$$

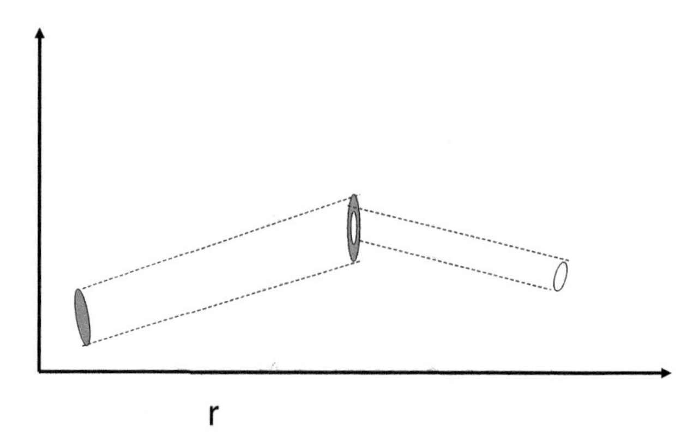

Fig. 4.8 A sketch of a simplified semiclassical model with the r being the translational degree of freedom and α the angle of the bound motion whose action is represented as circular cuts of the reaction channel. The distortions occurring during the advance inside the tube together with the differences in the cuts generate a mismatch between the asymptotic fluxes and the one at the junction between the reactant (LHS) and the product (RHS) half-channels

$$H = \frac{1}{2\mu}p^2 + H_o(\Delta) + V(\Delta, \alpha, r) \tag{4.62}$$

where α is the angle of the action-angle (Δ, α) pair of conjugated variables for the internal motion (p, r). Following Ref. [66], the wavefunction may be expressed in the JWKB form as

$$\psi(r, \alpha) = \exp\left[i W(\alpha, r/\hbar)\right] \tag{4.63}$$

and the Hamiltonian has the Taylor expansion

$$H_o(\Delta) + V(\Delta, \alpha, r) = \sum_k h_k(r, \alpha) I^k. \tag{4.64}$$

Accordingly also W can be expanded in Taylor series $W = W_o + \sum_k \hbar^k W_k$. In the $\hbar^0$ order, the wavefunction may be expressed as

$$\psi(r, \alpha) = (2\pi v)^{-1/2}\left[(\partial\alpha_1/\partial\alpha)_{n_1}\right]^{1/2} \exp\left[i W_o(\alpha, r)/\hbar\right] \tag{4.65}$$

or by transforming Eq. (4.65) in the asymptotic regions from (α, r) to a new $(\overline{\alpha}, t)$ representation ($\overline{\alpha} = \alpha - \omega t = \alpha - \omega\mu r/p$ with ω being $\partial H_o/\partial I$ and $\overline{\alpha}$ a constant) that by the comparison with the standard asymptotic form

$$\psi_{n_1 E}(\overline{\alpha}, t) \overset{t\to\infty}{\sim} (2\pi)^{-1/2} \sum_{n_2} S_{n_1 n_2} \exp\left[i n_2 \overline{\alpha} - i E t/\hbar\right] \tag{4.66}$$

that is v independent and allows the evaluation of $S_{n_1 n_2}$

$$S_{n_1 n_2} = \frac{1}{2\pi} \int_0^{2\pi} \left(\frac{\partial \overline{\alpha}}{\partial \overline{\alpha}_1} \right)_{n_1}^{1/2} \exp\left[i \Delta_{n_1 n_2}(\overline{\alpha}_1) \right] d\overline{\alpha}_1 \quad (4.67)$$

in which

$$\Delta_{n_1 n_2}(\overline{\alpha}_1) = [n(n_1, \overline{\alpha}_1) - n_2] \overline{\alpha}(n_1, \overline{\alpha}_1) - \int_{n_1}^{n(n_1, \overline{\alpha}_1) \alpha dn} - \int_{k_1}^{k(n_1, \overline{\alpha}_1) r dk} \quad (4.68)$$

that can be expressed also in terms of the cartesian coordinates using the proper classical generator $F_2(x_i, n_i)$ of the related transformation (see Appendix C of [3]).

In general, the probability amplitude $P_{1 \to 2}$ for transitions from the initial bound state 1 to the final bound state 2 (the square modulus of the related **S** matrix element) reads

$$P_{1 \to 2} = \sum_{roots} \int dx_1 \int dx_2 \psi_2^*(x_2) \psi_1(x_1) \left[(2\pi i \hbar)^F \left| \frac{\partial x_2}{\partial P_1} \right| \right]^{-1/2} e^{i S_i(x_2, x_1)/\hbar}, \quad (4.69)$$

where F is the number of degrees of freedom, P_i is the momentum of the ith state, and $S_i(x_2, x_1)$ is the classical action associated to the root trajectory i. This expression requires the search of all the root trajectories (the set of trajectories connecting "exactly"(though numerically) state 1 and 2). Closed-form solutions to such integrals were given by different authors using uniform mappings of the classical action associated with the root trajectories of simple models (Bessel, Airy, forced Harmonic oscillator, etc) so as to include the cases in which some root trajectories may coalesce (and therefore interfere) [3].

A limit of such formulation is given by the iterative nature of the search for root trajectories that disrupts concurrency and therefore prevents the distribution of the calculations. This need is avoided by reformulating the S matrix elements in terms of the initial conditions (initial value representation (IVR)) as follows:

$$S_{1 \to 2}^{IVR}(E) = -e^{-i(k_1 R_1 + k_2 R_2)} \int dP_{r_o} \int dr_o \int dP_{R_o} \left[\left| \frac{\partial(r_t, R_t)}{\partial P_{r_o}, P_{R_o}} \right| (2\pi i \hbar)^{-F} \right]^{1/2}$$
$$e^{-i[E_t + S_t(P_{r_o}, r_o, P_{R_o} +, R_o)]/\hbar} \psi_2(r_t) \psi_1(r_o) \hbar (k_2 k_1)^{1/2} / P_{R_t} \quad (4.70)$$

avoiding so far the need for singling out root trajectories and the division by the Jacobian determinant (now moved from the denominator to the numerator). Such formulation bears the advantage of allowing a quantum-like formulation of the state-to-state S matrix by using the whole outcome of the trajectory calculations. In practice, the rate coefficient $k(T)$ can be expressed in terms of the flux–flux correlation function $C_{ff}(t)$

$$k(T) = \frac{1}{Q_{trans}(T) Q_{rot}(T)} \int_o^{\infty} C_{ff}(t) \quad (4.71)$$

in which $Q_{trans}(T)$ and $Q_{rot}(T)$ are the translational and rotational partition functions of the system while the correlation function C_{ff} is defined as $C_{ff}(t) = R_{ff}(t)C_{ff}(0)$ that is in terms of a static factor $C_{ff}(0)$ (that can be evaluated as a partition function in the asymptotic region then mapped into the interaction region) and a dynamic one $R_{ff}(t)$ (that can be evaluated by replacing the exact time evolution propagator with the Herman and Kluk (HK) one). In order to evaluate $C_{ff}(t)$ therefore one can define a coordinate x along which locates a surface $s(x)$ separating the reactant configurations from the product ones.

4.4 Basic Features of Atom–Diatom Reactions

4.4.1 Energy Dependence of the Detailed Probabilities

As already mentioned, the possibility of carrying out accurate quantum calculations of the **S** matrix elements of the atom–diatom reactions provides us with a picture of the corresponding elementary processes that can hardly be paralleled by the experiment in terms of details. The key feature of such calculations is that they can be extended to conditions in which either the experiment cannot be performed or its outcomes are mixed with those of other intervening processes. Moreover, the computational study has the advantage of making explicit all the relationships and interactions between the intervening particles and the produced results allowing so far a rationalization of the computed outcomes.

In order to illustrate some important features of the atom–diatom reactions, we consider here the outcomes of reduced dimensionality calculations performed on a few emblematic cases of this type of systems. For this purpose, we consider here first the $N + N_2$ (nitrogen atom nitrogen molecule) system. Typically, nitrogen is quite inactive at normal conditions and scarce experimental information is available about related processes. However, the $N + N_2(v, j) \rightarrow N + N_2(v', j')$ reaction for a large variety of vibrational and rotational numbers are the dominant processes in the modeling of reentering spacecrafts and some plasmas. In these processes, the temperature is large and involves so far reactive transitions from/to a large number of vibrotational states and collision energies (the corresponding temperatures around reentering spacecrafts can be, for example, as high as several ten thousand degrees). Electronic structure calculations of the N_3 system have been performed in the past and a potential energy surface of the LEPS type has been fitted to the calculated points [67]. As a matter of fact, the contours of the LEPS plotted in Fig. 4.1 for the collinear geometry ($\Theta = 180°$) show that the reaction channel is indeed symmetric (entrance and exit have the same shape and contours) and that the barrier separating the entrance and exit channel is right in the middle (more recent calculations and fitting show that the barrier hosts on its top a little well). The dependence of the height of the MEP on the angle expressing its evolution from the reactant to the product channel (whose maximum is the barrier to reaction) for different values of Θ for

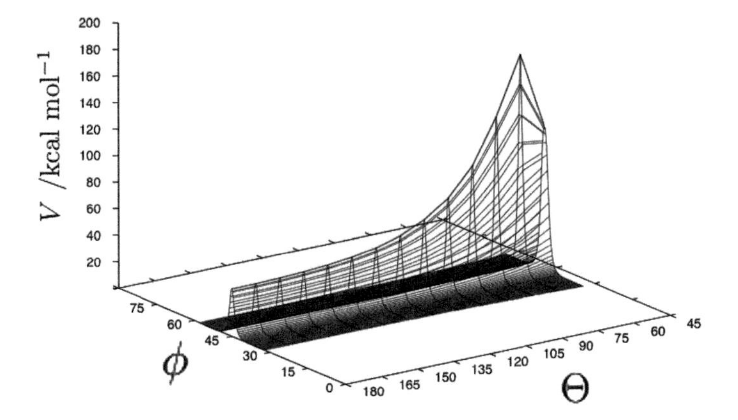

Fig. 4.9 Dependence of the minimum energy path plotted as a function of the angle expressing its evolution from the reactant to the product channel (whose maximum is the barrier to reaction) at different values of Θ for the N + N_2 LEPS

the considered LEPS is plotted in Fig. 4.9 and shows to be minimum for collinear encounters and maximum for the perpendicular ones (T-like, $\Theta = 90°$).

As a result of the fact that the PES exhibits a barrier to reaction, one intuitively expects that the reactive probability of the N + N_2 reaction exhibits a threshold in the dependence on energy. This is, indeed, true and this is what has been found in the early years of dynamical studies using classical trajectories. This is also confirmed by the curves of vibrational state specific $N + N_2(v) \rightarrow N + N_2$ quantum reactive probability when plotted as a function of the translational energy of the reactants at different vibrational quantum numbers v (see Fig. 4.10). As a matter of fact, the figure shows that reactive probabilities have an increasing trend as the collision energy increases confirming the importance of the translational degree of freedom in promoting reactivity. However, the various probability plots computed at different initial vibrational number show to be increasingly more reactive and confirm, therefore, also the effectiveness of vibrational energy in promoting reactivity. As a matter of fact, while the reactive probability of the ground vibrational level shows a threshold of about $E_{tr} = 1.42$ eV the value of the threshold lowers in energy as the reactants get more vibrationally excited. This effect prompts the question of which degree of freedom is more effective in promoting reactivity. This question has been tackled from the very beginning of reactive dynamics studies and was mainly associated with the position of the saddle to reaction: early saddle (located in the entrance channel) systems are more affected by collision energy, late saddle (located in the exit channel) systems are more affected by vibrational energy [68]. Quantum calculations confirm the effect and, obviously, more accurately and quantitatively treat their interplay.

To rationalize this in a more quantitative way it is helpful Fig. 4.11 in which the reactive probabilities of the N + N_2 reaction are plotted as a function of total energy (E). The figure shows clearly that in going from $v = 0$ to $v = 1$ the additional energy

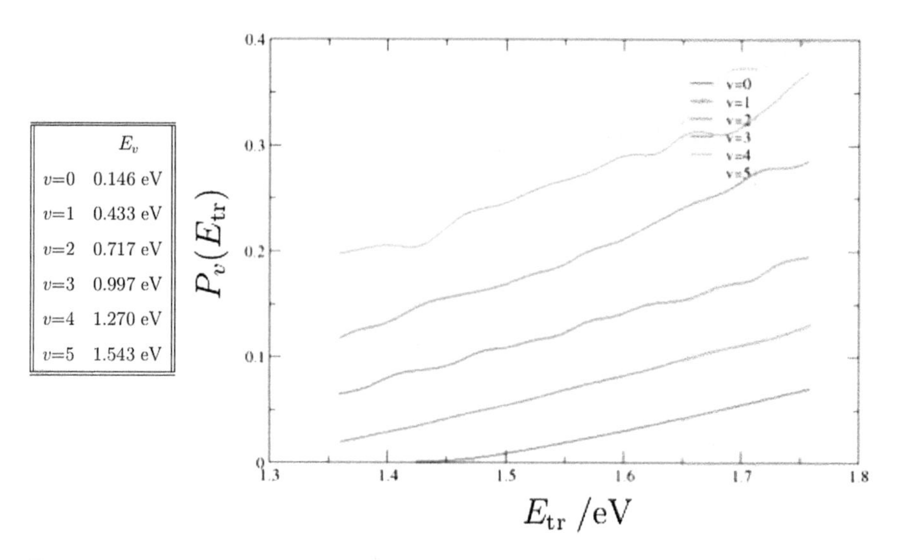

Fig. 4.10 Plot of the $N + N_2$ reactive probability (from $v = 0$ to $v = 5$) as a function of the collision energy

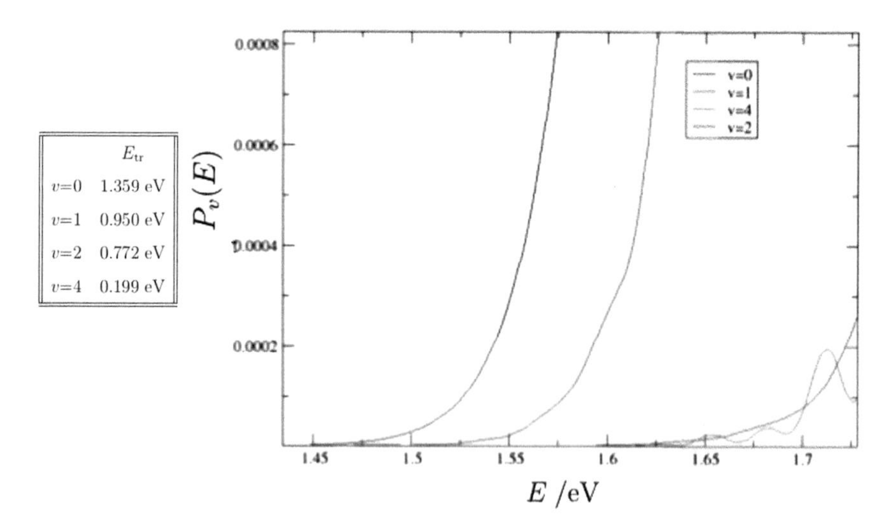

Fig. 4.11 Plot of the $N + N_2$ reactive probability (from $v = 0$, 1, 2 and 4) as a function of the total energy

poured into the system by vibrationally exciting the reactants is scarcely effective in promoting reactivity and does not lower the reactive threshold whereas the conversion trade-off between collision and vibrational energy does not affect the location of the threshold.

It is worth pointing out here that a simple qualitative interpretation of the effect of vibration and translation is that while excess energy allocated to translation helps surmounting the energy barrier located in the entrance channel, excess energy allocated to vibration helps surmounting (due to the "sideway" nature of the corresponding degree of freedom) the energy barrier located in the exit channel. More illustrative examples of such mechanisms will be given later together with the evidence that an excess of vibration or translation energy may end up preventing reaction. In some cases, not discussed here, however, the cooperative interplay of energy between the two degrees of freedom may help reaction (e.g., an attack from the side of repulsive interaction may slow down the collision and allow vibration to become effective once the attacked molecule has slightly rotated so as to offer its attractive side to the collision partner).

4.4.2 *Quantum Effects*

The smoothness of the quantum reactive probability plots of the $N + N_2$ system should not be taken as a general feature of chemical reactions. There are cases, in fact, in which the quantum reactive probability plots are much quite structured and it would be interesting to trace back the origin of such structures. It is interesting, in fact, to find out the nature of the underlying dynamical effect. This is indeed the case of the $H + H_2$ reaction in which the "lightness" of the collision partners exalts the quantum nature of the collision process. This makes the hydrogen atom–hydrogen molecule reaction the most popular benchmark for the calculation of quantum effects in atom–diatom collisions [69]. For illustrative purposes, the quantum reactive vibrational state-to-state probabilities of the collinear $H + H_2(v) \rightarrow H + H_2(v')$ case computed using the time-independent APH method are plotted in Figs. 4.12 and 4.13, respectively, as solid lines.

In Figs. 4.12 and 4.13 also, the IVR-SC values are plotted providing a qualitative comparison of the accuracy of the semiclassical treatment for both reactive and nonreactive processes. Systematic applications of the IVR-SC method have been made to the $N + N_2$ collisions [70].

In order to help the rationalization of the dynamical effects in elementary reactions let us consider the reaction Li + FH. This system bears the feature of being made of three different atoms (H that is by definition the lightest stable atom and Li and F which are respectively 7 and 19 times heavier than H and can, therefore, be considered both heavy). Such system has reaction channels largely differing from those of $H + H_2$. As shown by the minimum energy path plotted in Fig. 4.14, the channel Li + FH $\rightarrow$ LiF + H that connects the reactants to the LiF product is slightly endoergic (about 0.15 eV). The MEP, in fact, while Li approaches FH forms a well (associated with a slightly bent LiFH triatom) before rising to a double barrier (associated with a tightly bent triatom) sandwiching a small well of about 0.05 eV. The other channel connecting the reactants to the LiH product is instead highly endoergic and is not considered here.

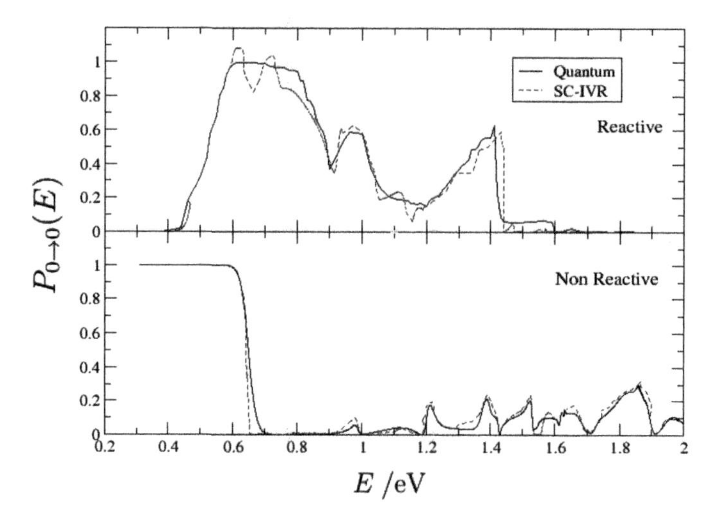

Fig. 4.12 Plot of the reactive probability for the $H + H_2(v = 0) \rightarrow H + H_2(v' = 0)$ reactive (upper panel) and nonreactive (lower panel) processes. Quantum values are given as solid lines, IVR-SC are given as dashed lines

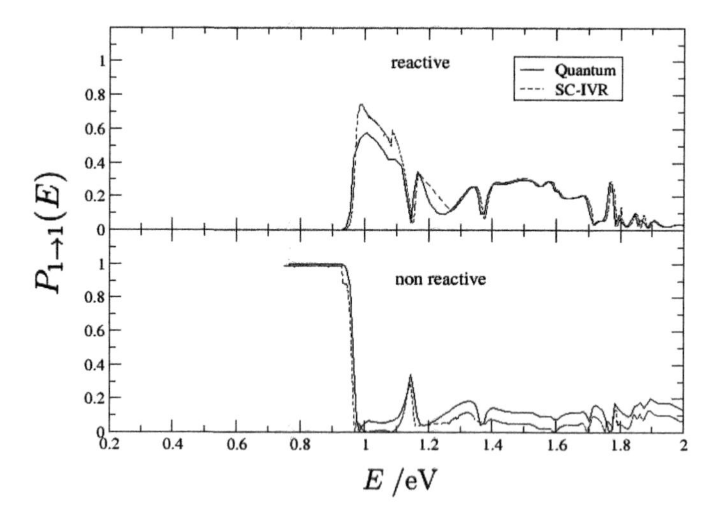

Fig. 4.13 Plot of the reactive probability for the $H + H_2(v = 1) \rightarrow H + H_2(v' = 1)$ reactive (upper panel) and nonreactive (lower panel) processes. Quantum values are given as solid lines, IVR-SC are given as dashed lines

The related adiabats are plotted in Fig. 4.15 and can be partitioned at long range in two subsets of which one is pretty flat at long and intermediate range and one drops earlier from higher values (the zero point energy of Li + FH is larger than that of H + LiF and falls while the system feels the entrance channel well. The adiabats and their avoided crossings provide the ground for rationalizing the main resonant features of the energy dependence of the probabilities plotted in Fig. 4.16.

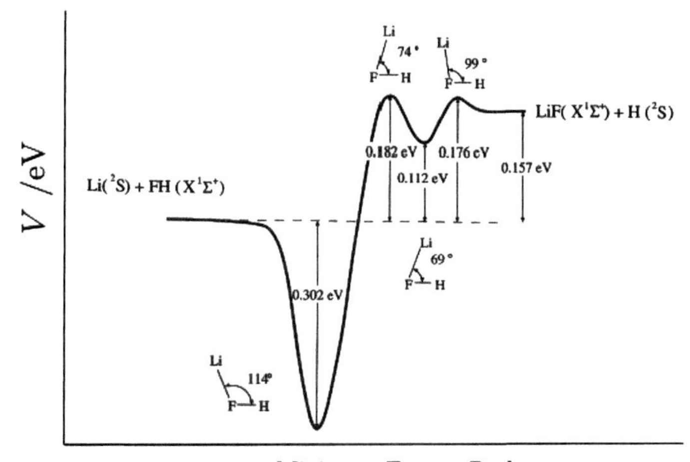

Fig. 4.14 Minimum Energy Path of the Li + FH → LiF + H reaction

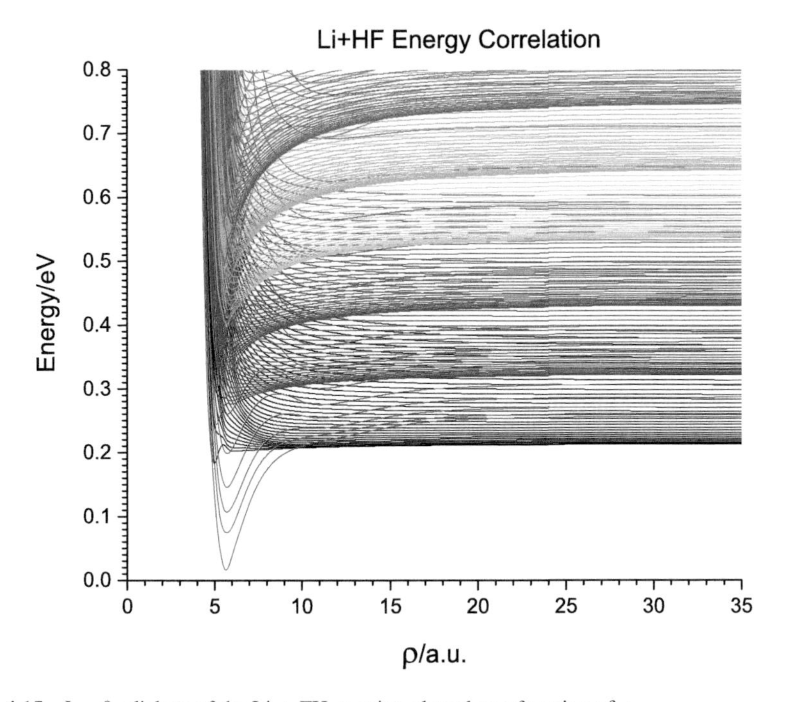

Fig. 4.15 $J = 0$ adiabats of the Li + FH reaction plotted as a function of ρ

In particular, the dense grid of resonances shown by the probabilities of Fig. 4.16 for $J = 0$ can be associated with the bound states supported by the wells formed by the adiabats as ρ varies. Particularly interesting is the fact that, as can be clearly seen

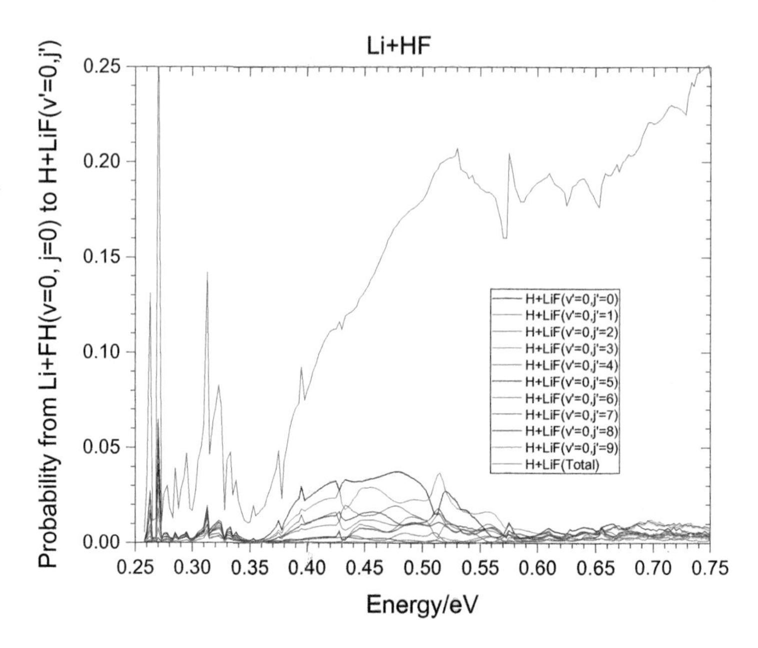

Fig. 4.16 State-to-state probabilities of the Li + FH reaction plotted as a function of energy for the null value of the total angular momentum

from Fig. 4.17, the fingerprint of the mentioned bound states is reproduced by all the plotted reactive state-to-state probabilities.

Above-commented findings for the state-to-state probabilities are corroborated by the fixed J probabilities plotted in Fig. 4.17 for increasing values of J. The plots show clearly that an increase in the total angular momentum leads not only to a shift in energy of the probability curves (as assumed by the already mentioned J-shifting approximation) but also to a smoothing of their threshold fine structure up to its almost complete disappearance.

4.4.3 Experimental Observables

As already mentioned, most often the ultimate goal of a theoretical investigation is the evaluation of hypothetical experimental observables measured under conditions inaccessible to the experiment once the theoretical procedure has been validated via a comparison of computed observables with measurements performed for tested conditions.

We consider here, as an example, the validation of the computational procedure carried out for the N + N_2 process by a comparison with limited experimental information.

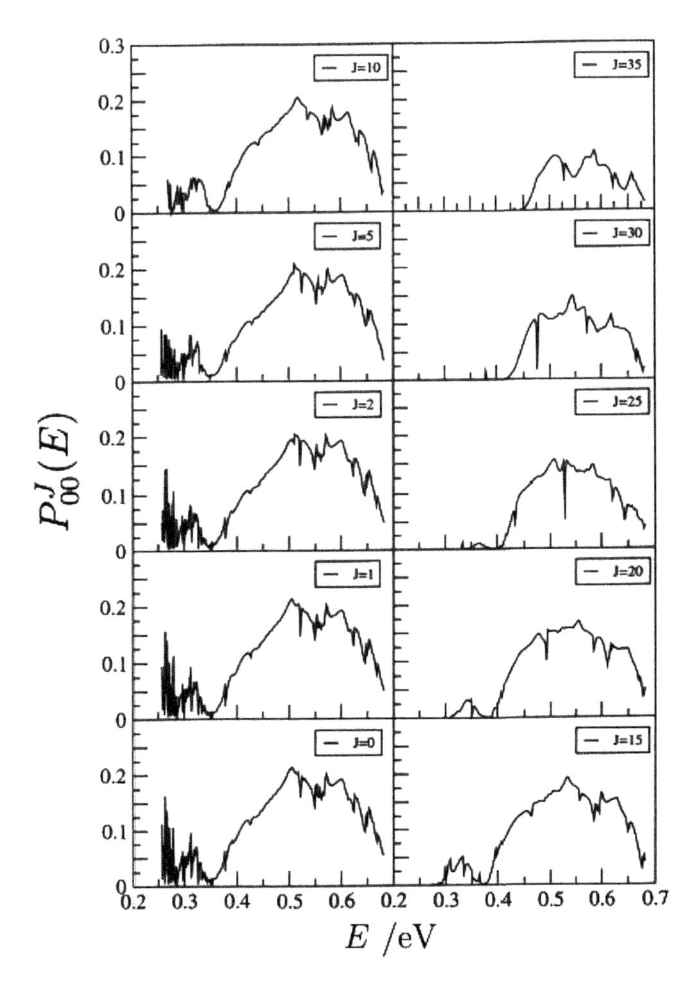

Fig. 4.17 The $j = 0, v = 0$ probability of the Li + FH reaction plotted as a function of total energy for increasing value of the total angular momentum

This is, indeed, the case shown in Fig. 4.18 for the thermal rate coefficient of N + N_2 computed using different methods and compared with values given in related experimental work. As can be clearly seen from the figure, the largest deviation from high level quantum calculations are for quasiclassical (QCT) results. A clear improvement is obtained when using the Reactive IOSA reduced dimensionality quantum and the semiclassical SC-IVR techniques. The lowest three curves are generated using the mentioned high level quantum techniques on slightly different PESs and confirm that quantum accurate values are on the average smaller. Such validation has allowed us to use the developed procedure with some confidence to assemble a database of vibrational state-to-state rate coefficients to be used by the European Space Agency for modeling the higher temperature environment of reentering spacecrafts.

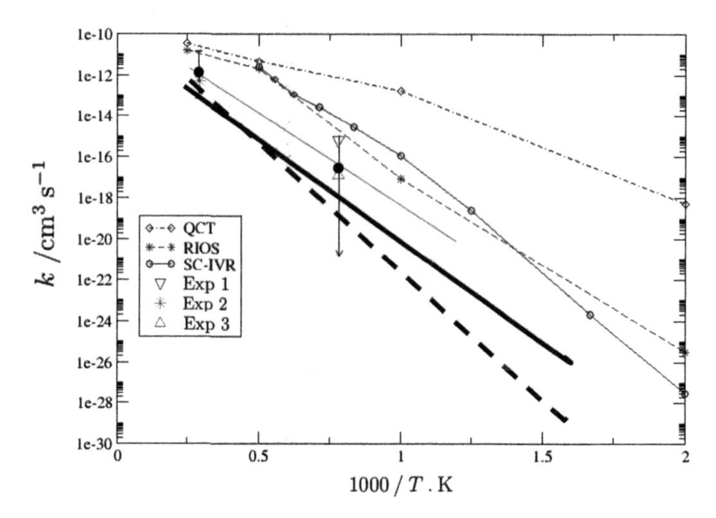

Fig. 4.18 Quasiclassical (diamonds connected by dashed line), Reactive IOSA (stars connected by long dashed line), SC-IVR (circles connected by solid line) values of the thermal rate coefficient of the N + N_2 reaction plotted as a function of the inverse temperature. For comparison, three experimental data (taken from Refs. [71–73]) and three quantum estimates (the lowest three lines) computed on slightly different PESs are also shown

Another example of comparison of quantum (in the case considered here using time-dependent techniques) full dimensional state-to-state probabilities (solid line) associated with reactive molecular collisions (collisional spectroscopy) for the HCl_2 exoergic reaction with the corresponding Reactive IOSA (dashed line) ones computed at different initial vibrational states and collision energy. The collisional (normalized) product vibrational distributions (PVD) given in Fig. 4.19 show the plots originating from the vibrational state ($v = 4$) and ($v = 5$) to vibrational state v'. The plotted PVDs exhibit a Frank–Condon-like shape (a sudden almost fully vibrational state-conserving transition) typical of highly exoergic reactions at both reactant vibrational numbers regardless of whether the full dimensional or the Reactive IOSA method is used. The only clear difference shows up in the fact that the full dimensional results exhibit a secondary (minor) peak at higher product vibrational number (7 and 8 respectively singling out a microscopic branching on different reactive paths to the same product).

4.4.4 *Periodic Orbits and Statistical Considerations*

The branching in various (microscopic and macroscopic) reaction paths is the fingerprint of the complexity of the reaction path followed by some collisions. The hydrogen atom + halogen molecule reaction discussed above provides further evi-

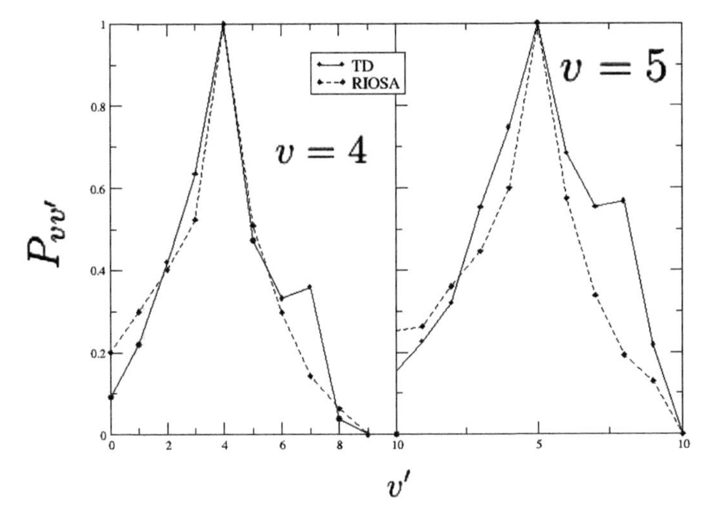

Fig. 4.19 Full quantum (solid line) and Reactive IOSA (dashed line) product vibrational distributions computed for the H + Cl_2 reaction at low collision energy

dence of the complexity of some trajectories due to the interplay of vibrational and translational energy as already singled out when investigating the access to the product channel by the N + N_2 reaction: an excess of both translational and/or vibrational energy may prevent reaction. For the H + Cl_2 reaction considered here, this occurs in spite of the fact that the H + Cl_2 MEP has only a small barrier in the entrance channel and above such small amount of energy would naturally drive reactants into the product channel.

Figure 4.20 shows, indeed, in the central panel, the bouncing back of the trajectory from the hard wall facing the entrance channel due to an excess of translational energy of the reactants. However, once the trajectory has bounced back, the full game is open again (in contradiction with the assumption of the transition state theory) and the interconversion of translational and vibrational energy may make the trajectory floats undecided on whether regressing to the entrance channel or head on into the product channel. The rationalization of this behavior can be obtained by locating on the considered PES the related PODS.[5] As shown by Fig. 4.21 there are PODS sitting on the reactant side (RHS) and PODS sitting on the product side (LHS) with increasing energy binding the regions of multiple crossing. They provide a guidance for the rationalization of the energy dependence of the reactive flux by counting the back and forth reflected trajectories (see Fig. 4.20).[6]

[5]PODS stands for periodic orbits dividing the surface (for surface here is meant an isoenergetic cut of the PES) a family of bound periodic trajectories (whose number and location depend on the energy at which the analysis is carried out) separating the phase space of reactants and products.

[6]For a detailed discussion on the use of PODS for defining the converging sequences of the number of odd crossing (forth) and even crossing (back) trajectories to improve the accuracy of the estimated reactive and non-probabilities see Ref. [3].

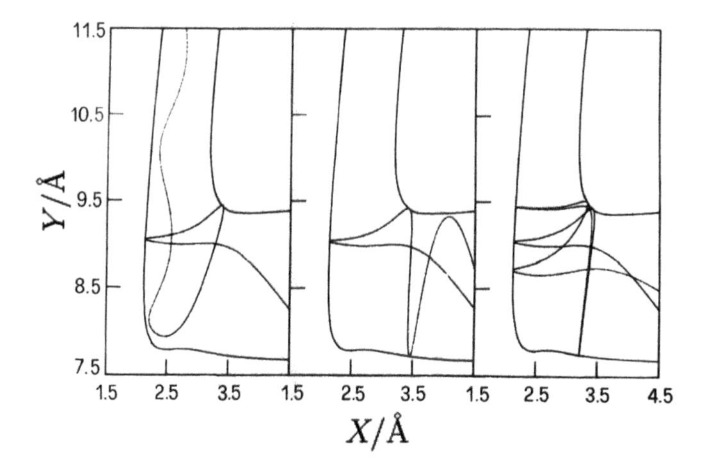

Fig. 4.20 Plot of trajectories with a high vibrational and translational energy content being reflected back by the repulsive wall facing the entrance channel. After bouncing back the trajectory can head to reactant asymptote (central panel) or to the product one (LHS panel). Bounced back trajectories, however, may still have a further complex fate by undergoing a few further forth and back oscillations before landing into its final asymptote (RHS panel)

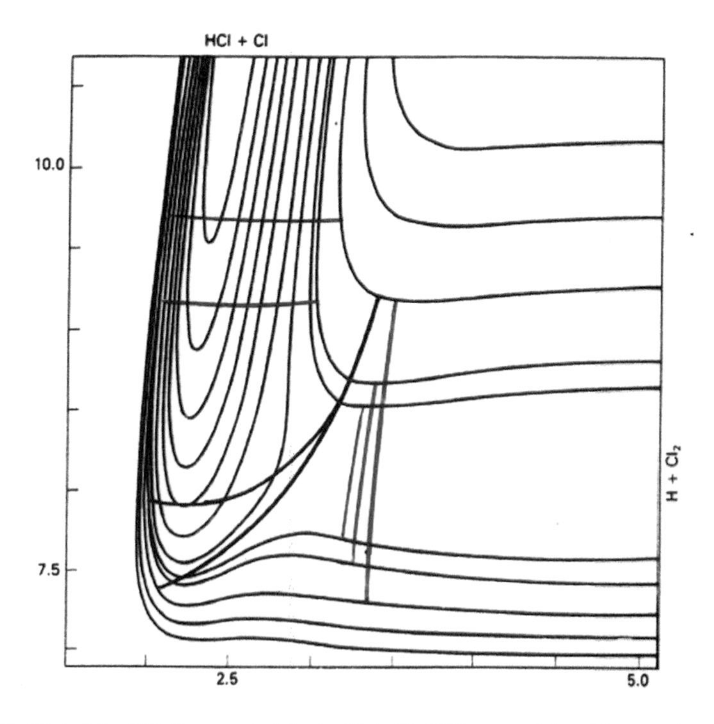

Fig. 4.21 Plot of the PODS computed on the H + Cl_2 PES mapped in arbitrary units on related energy contours

These considerations are useful when dealing with multidimensional systems and long lasting events to the end of estimating the error made by truncating the time integration on the accuracy of the computed reactive probability.

4.4.5 *The Last Mile to the Experiment*

The reactions already mentioned in this chapter

(a) $N + N_2 \rightarrow N_2 + N$
(b) $H + H_2 \rightarrow H_2 + H$
(c) $H + Cl_2 \rightarrow HCl + H$
(d) $Li + FH \rightarrow LiF + H$

are typical prototypes of atom–diatom reactive systems exhibiting in a different fashion-related distinctive features of related experimental data.

In particular, Reaction (a) is a typical heavy heavy-heavy (HHH) atom symmetric system with a dominant collinear minimum energy path. We have already seen, in the previous subsections, that the midpoint location of the barrier to reaction of this system leads to an even efficacy of translation and vibrational energy in promoting reactivity. At the same time, the isoenergicity of the system once past the threshold allows a non-negligible back reflection at certain energies that, however, due to the heavy masses involved do not show up as sharp spikes. They rather show up as shallow undulations which are completely smoothed in the highly averaged structure of the rate coefficient when plotted as a function of temperature (as shown in Fig. 4.18).

As to Reaction (b) (that is a typical light light-light (LLL) atom symmetric system also showing a dominant collinear minimum energy path), the light masses involved lead to a fine structure of the probability when plotted as a function of energy (as shown in Figs. 4.12 and 4.13). This is made possible by the frequent bouncing back and forth of the intermediate particle within the small skewed angle of the reaction channel for these systems.

Completely different is the behavior of Reaction (c) that is a typical light heavy-heavy (LHH) asymmetric exoergic system with a dominant collinear minimum energy path and an early barrier. The early location of the barrier enables collision energy to play a dominant role in promoting reaction. This agrees with the fact that a large excess of translational energy may cause the incoming particle to be reflected back so fast (short interaction time) as from a hard repulsive wall.

Another typical characteristic of the LHH character of this reaction and of its strong exoergicity is the typical Franck–Condon like nature of the product vibrational distributions that have a single peak at the product state energetically equivalent to the v state of the reactant (if $v = 0$) and are bimodal if $v = 1$.

Finally, Reaction (d), that is a typical light heavy-light (LHL) asymmetric endoergic system with a dominant noncollinear MEP, shows in the density of its adiabats both the light nature of the entrance channel diatom HF and the heavy nature of its exit channel diatom LiF. A peculiarity of this reaction is the fact that the third channel leading to LiH is highly endoergic (and therefore closed at thermal energies) and its behavior is exothermic like because of the much larger zero point energy of the reactant HF with respect to that of the product LiF to compensate for the endoergicity of the process.

As already mentioned, the above-described dynamical studies do not only offer a rationale for a more complete understanding of the mechanisms of elementary chemical processes but they can also offer the numerical procedures able to cover the last mile to the calculation of the signal measured by beam experiments (see next chapter for a clear example). They are, in fact, able to estimate on an *ab initio* fashion the signal of the experimental apparatus without making model assumptions provided that its geometry is fully specified.

In particular, for Reaction (d), the **S** matrix elements ($S^J_{vjl,v'j'l'}$) obtained, as already discussed, by projecting the asymptotic value of the propagated wavefunctions of the investigated atom–diatom system onto the corresponding final states of the channel of interest, bear the information necessary to construct the related measured integral cross section (the fixed energy E state vj to state $v'j'$ partial (fixed J) probabilities $P^J_{vj,v'j'}(E)$ obtained by squaring the $S^J_{vjl,v'j'l'}$ elements and properly summing them over l and l') allow, in fact, to evaluate the corresponding integral cross section

$$\sigma_{vj,v'j'}(E) = \frac{\pi}{k^2_{vj}} P_{vj,v'j'}(E) = \frac{\pi}{k^2_{vj}} \sum_J (2J+1) P^J_{vj,v'j'}(E). \qquad (4.72)$$

and have been used to cover the last mile aimed at reproducing the intensity of the product beam detected by the experimental apparatus.

4.5 Problems

4.5.1 *Qualitative Problems*

1. **Open Channels for** $H + H_2$: How many open rovibrational channels are there for scattering energies at $E = 0.7, 0.9, 1.1, 1.3, 1.5\,\text{eV}$ for the $H + H_2$ system assuming $J = 0$?

2. **Open Channels for** $F + H_2$: How many open rovibrational channels are there for scattering energies at $E = 0.7, 0.9, 1.1, 1.3, 1.5\,\text{eV}$ for the $F + H_2$ system assuming $J = 0$?
3. **Coupled Channels**: Consider the $H + H_2$ system. You have decided to include the lowest nine values of the rotational angular momentum $j = 0, 1 \ldots 8$ for each of lowest four vibrational states $\nu = 0, 1, 2, 3$. How many coupled channels will be used for $J = 0$, $J = 4$, $J = 8$, $J = 12$, $J = 16$, and $J = 100$. How many of these coupled channels will be open at $E = 0.7, 0.9, 1.1, 1.3, 1.5\,\text{eV}$?

4.5.2 *Quantitative Problems*

1. **ABC or APH3D**: Use the APH3D or ABC (http://ccp6.ac.uk/downloads.htm) programs to calculate the reaction probabilities for H_3 and $F + H_2$ for $J = 0$ and $J = 1$. Use total energies in the range $E = 0.3\,\text{eV}$ to $E = 1.5\,\text{eV}$. Do your answers agree with published results?
2. **TD_APH3D**: Use the TD_APH3D program to calculate the reaction probabilities for H_3 and $F + H_2$ for $J = 0$. Use total energies in the range $E = 0.3\,\text{eV}$ to $E = 1.5\,\text{eV}$. Do your answers agree with published results?
3. **Infinite Order Sudden Approximation—IOSSUD**: The infinite order sudden approximation is quite useful for calculating approximate inelastic nonreactive results. This approximation is valid when the total energy is large compared to the rotational energy spacing. It is equivalent to holding the orientation angle γ fixed between the atom and diatom during the collision. That is the molecule does not significantly rotate during the collision process. Use the JWKB phase shift program to calculate the phase shifts $\eta_l(\gamma)$ at 11 scattering angles for the $He + CO_2$ molecule assuming it is a rigid rotor. The orbital angular momentum quantum number should vary from 0 to 100 for each angle. Then calculate the integrated cross section by averaging over the orientation angles

$$\sigma = \frac{1}{2}\int_{-1}^{1} \sigma(\gamma)\, d\cos(\gamma), \tag{4.73}$$

where

$$\sigma(\gamma) = \frac{4\pi}{k^2}\sum_{l=0}^{l=100} (2l + 1)\sin^2\delta_l(\gamma) \tag{4.74}$$

The total differential cross can also be calculated as

$$I(\theta) = \frac{1}{2}\int_{-1}^{1} I(\gamma)\, d\cos(\gamma) \tag{4.75}$$

By the way, state-to-state differential cross sections, generalized cross sections, as well as bulk properties such as spectral line broadening, diffusion, viscosity, and virial coefficient can be calculated as well.

4. **Hyperangular bound states**: It is instructive to calculate the bound state energies of collinear H_3 atoms at various values of ρ. The eigenfunctions associated with these energies would be used in an adiabatic basis expansion of the scattering wavefunction. Use the PKH3 potential energy surface to generate the hyperangular potentials at several ρ ranging from $\rho = 2.5a_0$ to $\rho = 6.0a_0$ and plot the potentials. Then calculate the lowest five bound states for each ρ. Plot the eigenvalues as a function of ρ. Interpret this energy correlation plot.

Chapter 5
Complex Reactive Applications: A Forward Look to Open Science

This chapter focuses on the problem of more complex systems starting from those made of few atoms (mainly four) to move toward those made of several atoms and/or molecules. It starts from discussing the related formulation of the interactions for increasingly complex systems and continues by defining the quantities to be computed, describing some of the associated computational techniques and selecting the observables to simulate. Further considerations are made on the techniques used to describe the dynamics of large systems. Finally, the impact of the evolution of computer hardware and software on the progress of collaborative molecular science simulations is discussed. References to the open science and service-oriented scenario in computational activities targeting areas of societal relevance are also given.

5.1 Toward More Complex Systems

5.1.1 Full Range Ab Initio PESs for Many-Body Systems

In the previous chapters, we progressed from the description of analytical and computational approaches typical of simple model treatments of two-body problems (for which we dealt with methods based on special functions and elementary programs) to treatments aimed at handling complex molecular systems simulations dealing at the same time with electronic structure and nuclei dynamics (for which we dealt with large linear algebra and integrodifferential equation procedures). The progress made in dealing with the mentioned problems has found fertile ground in the fact that in the last 30 years, computer architectures have undergone a dramatic evolution from

A. Laganà and G. A. Parker (eds.), *Chemical Reactions*, Theoretical Chemistry and Computational Modelling, https://doi.org/10.1007/978-3-319-62356-6_5

single processor mainframes to multiprocessor high-performance compute (HPC) platforms.[1]

The libraries of ab initio electronic structure packages are numerous and a list of the most popular of them (together with a short description and some references) can be found on the web like in Ref. [17]:

NB-MCTDH [74] a multiconfigurational time-dependent Hartree program (MCTDH) for calculating bound states of a generalized N-body system including non-Born–Oppenheimer treatments.

MOLCAS [75] an ab initio quantum chemistry package for the multiconfigurational calculation of the electronic structure of molecules consisting of atoms from most of the periodic table with applications typically connected with the treatment of highly degenerate states (www.molcas.org).

MOLPRO [76] a system of ab initio quantum chemistry programs for molecular electronic structure calculations aimed at performing highly accurate computations by an extensive treatment of the electronic correlation problem through multiconfiguration-reference CI, coupled cluster and associated methods.

GAMESS-US [77] an ab initio electronic structure molecular quantum chemistry package that calculates potential energy values for moderately large molecular systems using direct and parallel techniques on appropriate hardware.

GAUSSIAN [78] a quantum chemistry package to calculate potential energy values for moderately large molecular systems performed using direct and parallel techniques on appropriate hardware.

NWChem [79] a highly scalable package for large scientific computational chemistry problems making efficient use of available parallel computing resources (from high-performance parallel supercomputers to conventional workstation clusters) to handle biomolecules, nanostructures, solid state using quantum and classical (in all combinations) approaches.

TURBOMOLE [80] a package designed for robust and fast quantum chemical applications performing ground state (Hartree–Fock, DFT, MP2, and CCSD(T)) and excited state (full RPA, TDDFT, CIS(D), CC2, ADC(2), ...) calculations at different levels. The package carries out also geometry optimizations, transition state searches, molecular dynamics calculations, various properties and spectra (IR, UV/Vis, Raman, and CD) determination. It adopts parallel version of coded fast and reliable approximations.

ORCA [81] a flexible, efficient and computational chemistry easy-to-use general purpose tool for quantum chemistry with specific emphasis on spectroscopic properties of open-shell molecules. The package makes use of a wide variety of standard quantum chemical treatments ranging from semiempirical techniques to DFT and single- and multireference correlated ab initio methods including environmental and

[1] High-performance computing (HPC) refers presently to machines exhibiting performances at the level of exascale based on massive parallel computing. Typically, HPC technologies require large investments for their establishing, specialized staff for their management, and highly skilled programmers for the exploitation of their performances.

relativistic effects. It is used also by experimental chemists, physicists, and biologist interested in rationalizing their measurements thanks to its user-friendly style.

HONDOPLUS [82] a modified version of the HONDO-v99.6 electronic structure program that includes solvation methods and other capabilities among which diabatization (by defining diabatic molecular orbitals (DMOs) to reformulate CASSCF or MC-QDPT wave functions) and new methods to calculate partial atomic charges, avoid intruder states, extend the basis sets, introduce user-defined density functionals, and improve portability (see http://t1.chem.umn.edu/hondoplus/).

CADPAC [83] a suite of programs for ab initio computational chemistry calculations. The package can perform SCF (RHF, ROHF, UHF and GRHF, analytic gradients and force constants (numerical for UHF), location of stationary points for all SCF types) Møller–Plesset and other correlated electronic structure (multipole moments, distributed multipole analysis, polarizabilities, magnetizabilities, NMR shielding constants, infrared intensities, Raman intensities, VCD intensities, frequency-dependent polarizabilities and hyperpolarizabilities, excitation energies by RPA method, dispersion coefficients, effects of external fields, field gradients, and lattices) calculations (see http://en.wikipedia.org/wiki/CADPAC).

All these packages (and several other more popular in specific research areas) are designed for computing ab initio potential energy values following a Born–Oppenheimer approximation. Most of them (either open source or commercial), however, are articulated into several separate programs and have been often developed over many years as a stratification of heterogenous pieces of software. This makes it difficult to further develop and restructure them for efficiently running on modern computer architectures. Moreover, most of them are mainly designed to calculate the electronic structure of molecular systems around equilibrium. This gives rise to discontinuity problems (like those associated with the convergence to the right electronic state when dealing with dynamics problems in which the partners are driven to explore large intramolecular distances).

5.1.2 Fitting PESs for Reactive and Nonreactive Channels

The problem of accurately determining reactive and nonreactive channels of a PES goes together with that of properly fitting a (large) set of calculated ab initio electronic structure values. Traditionally, this has involved, in particular, the strong interaction regions in which the exchange of energy and matter between atoms and molecules occurs. This has been discussed already in Chap. 3 where, in the spirit of the MPE (many-process expansion) approach, specific attention was paid to the processes in which the structured nature of the collision partners is responsible for an intriguing pattern of energy exchanges (including the distortion of molecular arrangements, their dissociation and reassociation, etc.) as well as long-range orientations and vector correlations regardless of whether the exchange of mass (reaction) occurs or not. In the reaction-oriented approach of the present book, strong emphasis is given to the role played by the long-range part of the interaction in approaching the strong interaction region leveraging on the fact that, at the energies important for a large variety of gas phase chemical processes, the system spends a significant fraction of time at large distances.

For most of the systems made of three, or a few more, bodies (the case of many more bodies will be considered separately later), the long-range overall potential V is usually partitioned into a strong interaction internal component (named $intra$) of the closely aggregated (bonded) atoms and a weaker (longer range) one (named $inter$) between the bonded bodies and the loosely interacting ones moving fairly free either in the entrance or in the exit channel

$$V = V_{intra} + V_{inter}. \tag{5.1}$$

V_{intra} is usually formulated around the equilibrium geometry of the bonded atoms (e.g., as a combination of Morse potentials) while V_{inter} is usually formulated in terms of a long-range two-body like "effective" interaction components (the van der Waals size repulsion plus dispersion attraction) V_{vdW} plus an electrostatic V_{elect} term [84] as follows:

$$V_{inter} = V_{vdW} + V_{elect}. \tag{5.2}$$

V_{vdW} can be expressed as a bond–bond pairwise interaction (more appropriate than the atom–atom ones) because it leverages on the additivity of the bond polarizability in contributing to the overall (molecular) one and accounts indirectly for three-body like effects [85]. V_{elect} is instead formulated as an electrostatic interaction associated with an anisotropic distribution of the molecular charge over the two interacting bodies (say molecule (or atom) a and molecule (or atom) b (sometimes labeled instead as 1 and 2)) that asymptotically tends to the permanent multipole – permanent multipole interaction.

Both V_{vdW} and V_{elect} are usually taken as functions of the intermolecular distance R between the centers of mass of molecule a and molecule b. For the simplest atom–diatom systems, the internuclear distances are used to formulate the strong interaction terms of the LEPS PES while the Jacobi coordinates are preferred for the V_{vdW} and V_{elect} interaction terms (both types of coordinates are illustrated in Fig. 4.2). As a matter of fact, the $N + N_2$ reaction already discussed in Chap. 3 is an appropriate example in order to illustrate how to exploit BO coordinates for three and more atom systems also for formulating both V_{vdW} and V_{elect} that are instead usually expressed as a function of the atom to the diatom center of mass distance R [39, 86].

A similar procedure has been adopted also for four N atom systems (in particular for the $N_2 + N_2$ diatom–diatom case). The Jacobi coordinates R, θ_1, and θ_2 formed by R with the internuclear vectors $\mathbf{r}_1$ and $\mathbf{r}_2$, respectively, and the angle Φ the dihedral angle formed by the planes $(R, \mathbf{r}_1)$ and $(R, \mathbf{r}_2)$ are illustrated in Fig. 5.1. The analysis of the interaction and the related fitting is performed by focusing on some representative configurations like the $(\theta_1, \theta_2, \Phi)$ = (90°, 90°, 0°), (90°, 90°, 90°), (90°, 0°, 0°), (0°, 90°, 0°), and (0°, 0°, 0°) ones.

The van der Waals term V_{vdW} is then formulated as an improved Lennard-Jones (ILJ) [87]

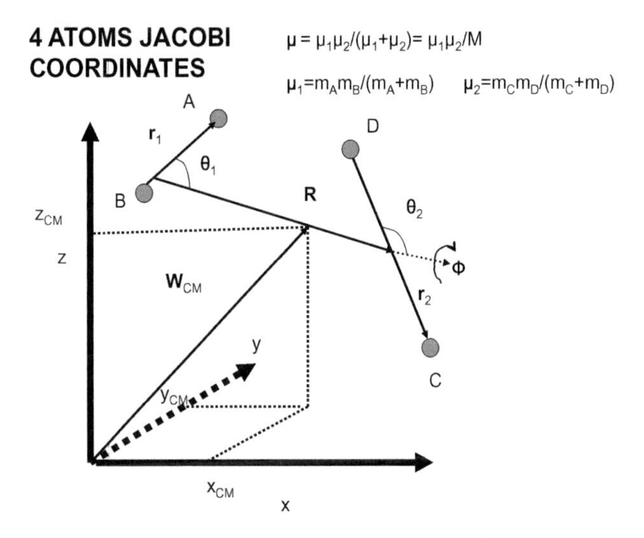

Fig. 5.1 One set of four-body Jacobi coordinates for the diatom–diatom case

$$V_{vdW}(R,\gamma) = \varepsilon(\gamma)\left[\frac{6}{n(x)-6}\left(\frac{1}{x}\right)^{n(x)} - \frac{n(x)}{n(x)-6}\left(\frac{1}{x}\right)^{6}\right] \tag{5.3}$$

with x being the reduced distance of the two bodies defined as

$$x = \frac{R}{R_m(\gamma)} \tag{5.4}$$

and γ denoting collectively the triplet of angles $(\theta_1, \theta_2, \Phi)$. In the same expression, ε and R_m are, respectively, the well depth of the interaction potential and the equilibrium value of R at each value of γ. Most often the van der Waals potential is also used in its reduced form

$$f(x) = \frac{V_{vdW}(R,\gamma)}{\varepsilon(\gamma)}. \tag{5.5}$$

The key feature of the ILJ functional form is the adoption of the additional (variable) exponential parameter n providing more flexibility than the usual Lennard-Jones (12, 6) [87] thanks to its dependence on both R and γ as

$$n(x) = \beta + 4.0\,x^2 \tag{5.6}$$

in which β is a parameter depending on the nature and the hardness of the interacting particles leading to a more realistic representation of both repulsion (first term in square brackets of Eq. 5.3) and attraction (second term in square brackets of Eq. 5.3). Additional flexibility is given to V_{vdW} by expanding ε and R_m in terms of the bipolar spherical harmonics in γ and taking the first terms. The value of coefficients of the expansion is then estimated from the abovementioned selected configurations further refined by computing high-level ab initio electronic structure. The comparison of computed cross sections and second virial coefficient values with beam experiments are also used for further refinements.

5.1.3 Four-Atom Many-Process Expansion

The additional flexibility built-in into the accurate description of the above discussed functional formulation of the PES of the four body systems sheds new light on the richness of its interaction components that have not yet found completely their way into LM-LS fitting methods for four-atom collisions. For the four-atom $N_2 + N_2$ processes the investigation of the interaction has focused initially only on the inelastic channel. Accordingly, the PES was initially formulated as a sum of the two N_2 intramolecular interactions and of an intermolecular component (i.e., that of two separated nitrogen molecules with their internuclear distances close to equilibrium) described in terms of isotropic and anisotropic contributions using expansions in spherical harmonics (see for example Refs. [90–95]) later formulated as a bond–bond pairwise additive interaction (see Ref. [96]). In order to overcome the bias of such formulation of the PES (that prevents the description of atom exchange processes and the fragmentation of one (or both) molecule(s)), extensive additional ab initio studies were performed for a wide set of molecular geometries aimed at obtaining a full-dimensional more rigorous description of the interaction governing the $N_2 +$ N_2 collisions. The first work along this direction was reported in Ref. [97]. A more complete effort to deal with the problem of describing the processes in which the initial N_2 molecules deform to the extent of reaching the situation of either forming new bonds or breaking old bonds and allowing the newly formed atoms and molecules to fly away, was made by the authors of Ref. [98] by carrying out *ab initio* computations for 16435 geometries describing nine $N_2 + N_2$ and three $N + N_3$ arrangements. Then, this set of potential energy values was fitted both to a polynomial of bond-order variables [98, 99] (Paukku PES) and to a statistically localized, permutationally invariant, local moving least squares interpolating function [100]. The same set of data was used to build a PES as a sum of bond, valence angle, torsion angle, van der Waals and Coulombic energy interaction terms among all atom pairs [101]. More recently, based on the same set of ab initio values, a new polynomial function (Bender PES) was proposed including in the definition of the bond-order variables a Gaussian contribution [102]. This feature provides the Bender PES with significant flexibility that improves the fitting quality of the ab initio values. What is more important, however, is that both the Paukku and the Bender PESs validly describe high energy processes (including dissociation) and consistently formulate (using polynomials) two-, three-, and four-body components (e.g., reproduce the double barrier structure of the triatomic $N + N_2$ subsystem when one nitrogen atom is displaced to very large distances).

Yet, the accuracy of the abovementioned PESs in describing the long-range interaction is far from being satisfactory. For this reason systematic attempts to provide a ROBO formulation of the four-body N_2+ N_2 processes PESs [88, 89] have been made by generalizing to four atoms the three-atom formulation. As discussed in Sect. 3.4.4, a way of simplifying LM-LS methods may consist in embodying in the procedure a criterion driving the selection of both **c** coefficients and **f** functions by inspiration from the process relevance proposed in the many-process expansion, MPE [34], method that articulates the PES in functional representations connecting different asymptotic arrangements going through different tightly bound many-body clusters (see Eq. 3.51). In this respect, the method tackles the problem of building the PES in terms of approaching paths converging toward a dynamically controlled sampling of the shorter range interactions. More in detail, in their present version the MPE potentials exploit the versatility of the BO variables and:

1. Express the **f** functions of the LM-LS methods as long-range formulations of both reactant and product channels of the considered processes in terms of either BO or SRBO variables replacing the Lennard-Jones-like ones.
2. Use the BO or SRBO variables determined in this way in order to build a polynomial representation of the PES in the intermediate (stable or pseudostable) region of the interaction.
3. Formulate the **c** coefficients in terms of angles of the (appropriately selected) involved ROBOs linking existing bonds to the newly (even if only locally or temporarily) formed ones. Criteria for driving their selection are based only on a tentative evaluation of the local importance of the subset of bonds undergoing distortion or (even if embryonal) formation.

To this end it is crucial to consider all the representations of the system (among which it can switch) including full 4 body aggregation, full 4 body fragmentation and all possible 3 and 2 body combinations. For all these processes one can always consider the 4 bodies in a sequence (say κ, λ, μ, ν) forming a dihedral angle ζ. Similarly to the atom–diatom case (see Fig. 5.2 one can define for four-atom systems its size variable ρ

$$\rho = [n_{\kappa\lambda}^2 + n_{\lambda\mu}^2 + n_{\mu\nu}^2]^{1/2}, \tag{5.7}$$

and the angles ϕ (for the first three bodies) and ϵ (for the second three bodies) as follows:

$$\phi = \arctan\left[\frac{n_{\mu\nu}}{n_{\kappa\lambda}}\right] \tag{5.8}$$

$$\epsilon = \tan^{-1}\left[\frac{n_{\lambda\mu}}{n_{\mu\nu}}\right] \tag{5.9}$$

by decoupling the whole process into two subprocesses. It is, in fact, simpler to consider separately the different subsystems as all spectators but one on which the

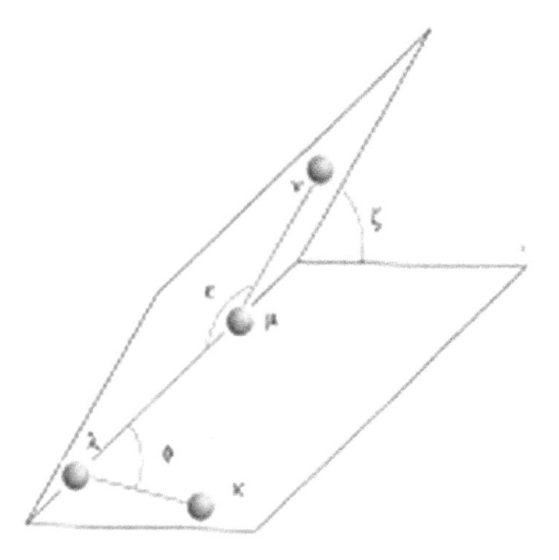

Fig. 5.2 A sketch of the four-body LAGROBO coordinates

transformation is performed in turn. One can also adopt the full four-body coupled formalism of diatom–diatom systems [36] and use ϕ, ϵ and the dihedral angle ζ altogether.

5.1.4 *Four-Atom Quantum and Quantum-Classical Dynamics*

The four-body (diatom–diatom) Jacobi coordinates discussed above lead to a formulation of the Hamiltonian operator as

$$\hat{H}_N = -\frac{\hbar^2}{2\mu}\frac{\partial^2}{\partial R^2} - \frac{\hbar^2}{2\mu_1}\frac{\partial^2}{\partial r_1^2} - \frac{\hbar^2}{2\mu_2}\frac{\partial^2}{\partial r_2^2} + \hat{T}_{ang} + V(R, r_1, r_2, \theta_1, \theta_2, \Phi)$$

with

$$\hat{T}_{ang} = \frac{(\hat{J} - \hat{j}_1 - \hat{j}_2)^2}{2\mu R^2} + \frac{\hat{j}_1^2}{2\mu_1 r_1^2} + \frac{\hat{j}_2^2}{2\mu_2 r_2^2} \tag{5.10}$$

with $\mu_1 = m_A m_B/(m_A + m_B)$, $\mu_2 = m_C m_D/(m_C + m_D)$, $\mu = \mu_1\mu_2/(\mu_1 + \mu_2)$, V the potential energy of the system. The procedure integrating the Schrödinger equation decomposes the wavefunction of the system into partial waves $\psi^{Jp}(R, r_1, r_2, \theta_1, \theta_2, \Phi, \alpha, \beta, \gamma)$ eigenfunctions of the total angular momentum J and parity p in terms of a radial component $\varphi^{Jp}_{j_2,\Omega_1\Omega}(R, r_1, r_2)$ and an angular one $G^{Jp}_{j_1,j_2,\Omega_1\Omega}(\vartheta_1, \vartheta_2, \Phi, \alpha, \beta, \gamma,)$ where α, β, γ are the Euler angles already defined in the previous chapter. For nonreactive systems the reactant coordinate formulation is kept all along the collision process. The integration of the Coupled-Channel, CC, equations resulting from the usual expansion in partial waves proceeds then through the steps already singled out in the previous chapter for the atom–diatom systems. This implies the carrying out of the propagation of the solution from the first sector to the last one, matching the calculated values with the asymptotic ones for all the necessary boundary conditions, working out of this comparison the value of the elements of the **S** matrix, performing the appropriate statistical averaging of the computed quantities to the end of calculating the corresponding observables. These steps, though conceptually identical to those of the atom–diatom case, become increasingly more difficult for heavier systems, more complex electronic structures and a larger coupling of the related calculations. For this purpose, large use of the package MOLSCATT [103] based on an expansion of the potential in terms of spherical harmonics, is made for nonreactive systems.

For reactive processes, it is usually preferred to adopt the hyperspherical formalism that defines the reaction coordinate as the hyperradius (regardless of the arrangement number and type of the atoms composing the molecular subsystem). However, the increasing difficulty of dealing with the fixed ρ surface functions (the eigenfunctions of the 3N-4 hyperangles) for N larger than 4, has till now made the generalized use of these coordinates computationally impractical for scattering calculations even if formally interesting.

For four-atom (diatom–diatom in our case) systems one can reduce the complexity of the problem using the Quantum-Classical (QC) Coupled-Channel method (see Refs. [104–106] for a more extended discussion) in which molecular vibrations are treated quantum-mechanically by integrating the related time-dependent Schrödinger equations for the N_2 and the O_2 molecules. On the contrary, translational and rotational degrees of freedom are treated classically by integrating the related classical Hamilton equations. The two subsystems, and the corresponding equations of motion, are dynamically coupled through the definition and calculation of a time-dependent "effective" Hamiltonian, of the Ehrenfest type, defined as the expectation value of the intermolecular interaction potential over $\Psi(r_a, r_b, t)$

$$H_{\text{eff}} = < \Psi(r_a, r_b, t) \mid V_{inter}(R(t)) \mid \Psi(r_a, r_b, t) > \tag{5.11}$$

where $V_{inter}(R(t))$ is the intermolecular interaction potential evaluated at each time step of the classical "mean" trajectory $R(t)$.

The time evolution of the total wave function is obtained by expanding $|\Psi(r_a, r_b, t) >$ over the manifold of the product, rotationally distorted, Morse wave functions of the two isolated molecules $\Phi_{v'_a}(r_a, t)$ and $\Phi_{v'_b}(r_b, t)$ as follows:

$$\Psi(r_a, r_b, t) = \sum_{v'_a, v'_b} \Phi_{v'_a}(r_a, t)\ \Phi_{v'_b}(r_b, t)\ e^{-i\frac{E_{v'_a}+E_{v'_b}}{\hbar}t}\ A_{v_a v_b \to v'_a v'_b}(t) \tag{5.12}$$

in which $A_{v_a v_b \to v'_a v'_b}(t)$ is the amplitude of the vibrational transition from v_a and v_b to v'_a and v'_b, $E_{v'_i}(t)$ is the eigenvalue of the v'_i Morse wavefunction $\Phi_{v'_i}(r_i, t)$ corrected by the Coriolis coupling terms $H_{v''_a v'_b}$

$$\Phi_{v_i'}(r_i, t) = \Phi^0_{v_i'}(r_i) + \sum_{v_i'' \neq v_i'} \Phi^0_{v_i''}(r_i) \frac{H_{v_i'' v_i'}}{E^0_{v_i'} - E^0_{v_i''}} \tag{5.13}$$

The second term in Eq. 5.13 represents the first-order centrifugal stretching contribution originating from the coupling of diatomic rotations and vibrations with $\Phi^0_{v_i'}$ and $E^0_{v_i'}$ being the eigenfunction and the eigenvalue, respectively, of the same Morse oscillator. In the same equation

$$H_{v_i'' v_i'} = -j_i^2(t) v_i''^{-1} (\overline{r}_i)^{-3} < \Phi^0_{v_i''} | r_i - \overline{r}_i | \Phi^0_{v_i'} > \tag{5.14}$$

with j_i being the rotational momentum of molecule i.

Thus, the Hamilton equations for the roto-translational motions are integrated self-consistently together with the Schrödinger equations of the vibrational amplitudes ($2N + 18$) coupled classical (18) and quantum ($2N$) equations with N being the total number of vibrational levels in the total wave function expansion). The number of vibrational levels, above and below the initial vibrational state of N_2 and O_2, included in the wave function expansion depends on the initial vibrational state of both molecules and on the impact kinetic energy. The higher the impact energy and the level of vibrational excitations of N_2 and O_2, the larger is the number of vibrational states required (and, therefore, the larger is the number of coupled wave equations to be solved). At the same time, the calculations need to be repeated for an ensemble of N_t trajectories large enough to sample adequately the range of initial values of both the diatomic rotational angular momentum (for both a and b, from 0 to $j_{a\max}$ and $j_{b\max}$, respectively) and the diatom–diatom orbiting angular momentum range (from 0 to $l_{\max}$).

Accordingly, the semiclassical cross section for the vibrational transition $v_a v_b \rightarrow v_a' v_b'$ (or $(v_a, v_b | v_a', v_b')$) is given by the following expression:

$$\sigma_{v_a v_b \rightarrow v_a' v_b'}(U) = \frac{\pi \hbar^6}{8 \mu I_a I_b (k_B T_0)^3}$$
$$\int_0^{l_{\max}} \mathrm{d}l \int_0^{j_{a\max}} \mathrm{d}j_a \int_0^{j_{b\max}} \mathrm{d}j_b \frac{[l j_a j_b]}{N_{v_a v_b}} \sum |A_{v_a v_b \rightarrow v_a' v_b'}|^2 \tag{5.15}$$

in which, as usual, k_B is the Boltzmann constant and μ and l are the reduced mass and the orbital angular momentum, respectively, of the colliding system. In the same equation I_i is the moment of inertia of molecule i, $[l j_a j_b] = (2 j_a + 1)(2 j_b + 1)$ $(2l + 1)$, U the classical part of the kinetic energy defined as $U = E_{kin} + E^a_{rot} + E^b_{rot}$ (with E_{kin} being the impact kinetic energy and E^i_{rot} the rotational energy) and T_0 a reference temperature (see Refs. [104, 105]) introduced in order to provide the proper dimensionality to the cross section formulation.

The state-to-state rate coefficients are then obtained by averaging over an initial Boltzmann distribution of kinetic and rotational energies

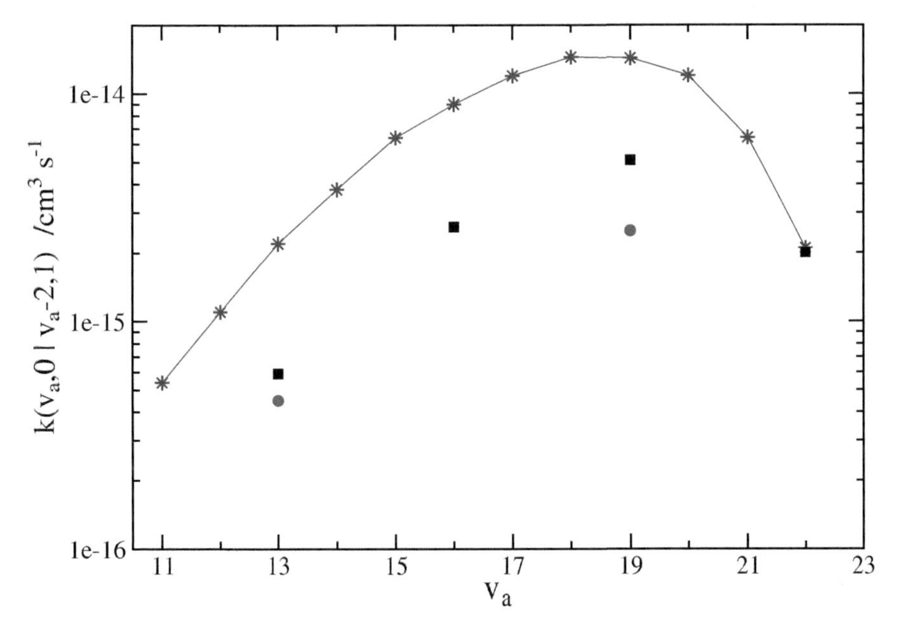

Fig. 5.3 Quasi-resonant rate coefficients of a diatom–diatom system computed on two different PESs (GB1 (solid squares) and MF (solid circles) as from Ref. [106]) plotted as a function of the initial vibrational number v_a and compared with experimental results (stars)

$$k_{v_a v_b \to v'_a v'_b}(T) = \sqrt{\frac{8k_B T}{\pi\mu}} \left(\frac{T_0}{T}\right)^3$$
$$\int_{\varepsilon_{min}}^{\infty} \sigma_{v_a v_b \to v'_a v'_b}(U) \exp\left[-\frac{\overline{U}}{k_B T}\right] d\left(\frac{\overline{U}}{k_B T}\right) \tag{5.16}$$

where $\overline{U}$ is the symmetrized effective energy ($\overline{U} = U + \frac{1}{2}\Delta E + \Delta E^2/16U$), with E being the total energy and $\Delta E = E_{v'_a} + E_{v'_b} - E_{v_a} - E_{v_b}$ being the exchanged energy (see Fig. 5.3 for a comparison with the experiment).

5.1.5 Last Mile Calculations for Crossed Beam Experiments

The problem of working out theoretically measurable quantities (in the "*last mile*" spirit of reproducing from "ab initio" the signal of the experimental apparatus) is of great importance for both theorists and experimentalists. Moreover, the above-discussed techniques offer suitable means for providing appropriate solutions to its quantitative solution. Yet, when reaction is involved, the procedures might end up to be quite cumbersome. On the contrary, for four (and even more) atoms studies, the simplicity of a trajectory treatment (that, though being a more recently developed computational procedure than those developed for electronic structure calculations, have already reached a similar robustness and popularity) can make the problem easily solvable. Using classical mechanics and cartesian coordinates, in fact, the time evolution of the system can be easily followed by integrating numerically the first-order ordinary differential Hamilton equations given in Eq. 1.29 for each ith body of the ensemble of the atoms to be treated once initial positions and momenta are defined. As already discussed for three atom systems, the possibility of both associating simple formulations of the reactant vibrational quantum states and sampling of the corresponding interatomic distances allows us to deal with four-atom systems (diatom–diatom in particular) and to determine related product (reactive and non reactive) states with a limited amount of extra work and little loss of accuracy.

A clear illustration of the potentialities of the collaborative molecular simulator named GEMS [107] (grid empowered molecular simulator whose detailed description will be given later) in supporting the rationalization of crossed molecular beams (CMB) experiments using classical trajectories is given below by discussing the investigation of the OH + CO reaction [108]. In this case, GEMS has been able to handle a reactive diatom–diatom study all the way through from first principle treatments to measured data reproduction (the so-called "*last mile*") of the intensity of the CO_2 product. In the CMB experiment the product number density $N_{lab}(\Theta, t)$ was measured in the laboratory frame as a function of the scattering angle, Θ, and time of flight, t, for OH and CO beams colliding with a narrow distribution of the reactant collision energy (E_c) around its nominal value and a single or a small set of reactant vibrotational states (v, j).

In most CMB experimental studies $N_{lab}(\Theta, t)$ is converted (using when necessary mechanistic model assumptions) into the center of mass, CM, differential cross section that corresponds to the CM product flux $ICM(\theta, u)$. In principle, it is possible to directly convert the measured $N_{lab}(\Theta, t)$ into $ICM(\theta, u)$ if a sufficiently fine grid of measured points is available and experimental measurements are sufficiently clean. However, the use of such direct inversion procedure is impracticable due to the finite resolution of experimental conditions (i.e., finite angular and velocity spread of the reactant beams and angular resolution of the detector). For this reason the analysis of the laboratory data is usually carried out by a forward convolution trial-and-error procedure, in which tentative (model) CM angular and velocity distributions are assumed, averaged and transformed to the *lab* distributions for comparison with the experimental data, until the best fit is achieved (with this procedure it is trivial to account for the averaging over the experimental conditions). Moreover, $ICM(\theta, u)$ is usually expressed in terms of a product angular (PAD) and a product translational energy (PTD) Distribution under the assumption that they are substantially uncoupled. The best-fit CM PAD and PTD are then usually compared with other measured data obtained under different experimental conditions and interpreted by associating their shapes with the reaction mechanism of some known models. The same quantities are then compared as well with the outcomes of theoretical calculations.

Yet, using the computational techniques illustrated so far $ICM(\theta, u)$ can be evaluated directly from first principles if one knows the experimental conditions and the machine geometry. This direct comparison of theoretically predicted laboratory distributions with experimental data avoids any arbitrariness of the models adopted when analyzing of crossed beam data.

This procedure provides the most accurate evaluation of the quality of the adopted ab initio PES and of the used dynamical treatment. It implies, however, the collaborative effort of research groups bearing different and complementary expertise. Namely, expertise in

(a) Ab initio calculation of molecular electronic structure,
(b) Fitting of the potential energy values to a functional form,
(c) Running of molecular collision dynamical calculations,
(d) Assembling experimental observables from elementary properties

need to be properly combined. In particular, the forward convolution step performing the appropriate averaging of detailed dynamical ab initio properties using a single stream procedure to evaluate from first principles the $N_{lab}(\Theta, t)$ distributions heavily relies on a tight collaboration between computational chemists and experimentalists. The measured $N_{lab}(\Theta, t)$ (solid circles) are compared in Fig. 5.4 with the computed (solid lines) *lab* PADs of CO_2 at $E_c = 14.1$ Kcal mole^{-1} for the Leiden (second panel from the top), YMS (second panel from the bottom) and LTSH (bottom panel) PESs obtained in this way [108].

Such computational procedure is, indeed, the one embodied into the so-called grid empowered molecular simulator (GEMS) [107] that is a workflow articulated into the following highly cooperative computational blocks:

- INTERACTION, for the generation and/or collection of single geometry ab initio Born–Oppenheimer electronic structure values related to the potential energy surface of the system of interest for the whole range of relevant geometries. For this purpose, one can choose among different suites of codes (including commercial packages). This block can be skipped when use is made of precalculated ab initio inputs.
- FITTING, for the use of either a global or a local interpolation of ab initio data belonging to the same electronically adiabatic surface in terms of one or more suitable functional forms (this block is skipped when an on the fly dynamical approach is considered or a force field is adopted or a suitable grid of values of the PES is already available from the literature). Most often the INTERACTION and FITTING are chained together (including sometimes also the next block DYNAMICS) and iteratively used to the end of better shaping some critical regions of the PES against information available from the literature.
- DYNAMICS, for the carrying out of dynamics calculations on the PES produced by the FITTING block (or on the individual potential energy values provided by ad hoc on the fly implemented ab initio packages). For few atom (at present only atom–diatom) systems, the dynamics problem can be dealt with by using full-dimensional quantum mechanics techniques and converging with total angular momentum. Approximate quantum, SC, QC, and QCT calculations can also be used.

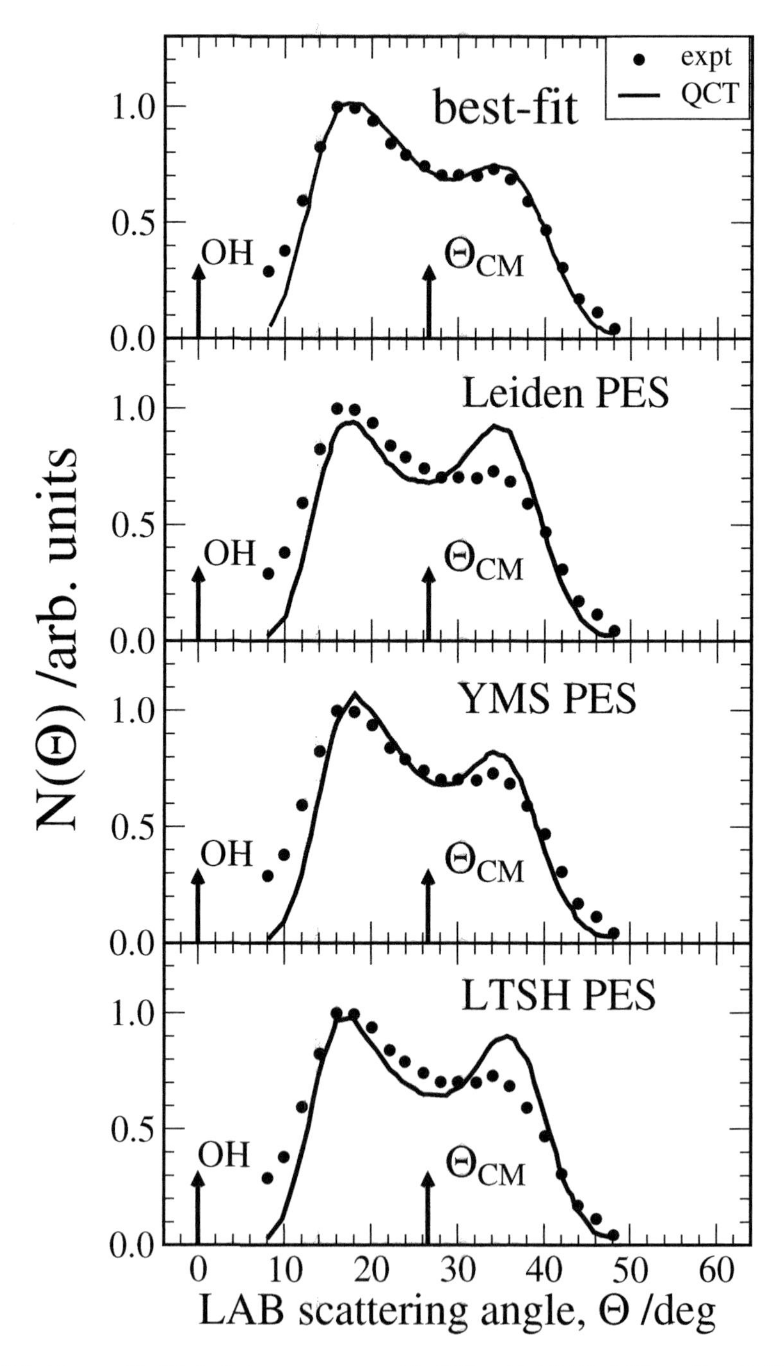

Fig. 5.4 Measured (solid circles) versus computed (solid lines) *lab* PAD of CO_2 at $E_c = 14.1$ Kcal mole^{-1} for the Leiden, YMS, and LTSH PESs (see Ref. [108]). Best fit to experimental results are shown in the top panel

- OBSERVABLES, for model and statistical treatments necessary to cover the last mile to measurable properties when the physical parameters of the experimental apparatus are available to allow the evaluation of the experimental signal (like when performing the conversion from CM quantities to the *lab* ones).

The GEMS scheme can be iteratively performed until theoretical outcomes agree with experimental data better than the error of the numerical approximations and/or the experimental uncertainties.

5.2 Large Systems Studies Using Classical Dynamics

5.2.1 Trajectory Studies for Many-Body Systems

The massive exploitation of the advanced features of parallel machines has impacted significantly on the performance of molecular science calculations especially for codes based on highly decoupled tasks. For these codes, in fact, data transfer is minimal and computations run as independent events resulting in a significant increase of the performances (especially if the memory used is small and the number crunching section of the procedure is large). This is the case, in fact, of many atoms trajectory (classical mechanics) codes. Trajectories, in fact, can be followed by integrating numerically the corresponding set of first-order ordinary differential Hamilton equations (see Eq. 1.29) or equivalent ones like the Newton's second law for which each particle has

$$\mathbf{F}_i = M_i \mathbf{a}_i \tag{5.17}$$

where $\mathbf{F}_i$ is the force acting on the particle i a mass point m_i and $\mathbf{a}_i$ is its acceleration. For the integration of Eq. 5.17 in the generic position vector $\mathbf{W}$ of its particles, the second order Verlet algorithm [109]

$$\mathbf{W}_i(t + \Delta t) \equiv -\mathbf{W}_i(t - \Delta t) + 2\mathbf{W}_i(t) + (\Delta t)^2 \frac{d^2}{dt^2}\mathbf{W}_i \tag{5.18}$$

is often used. The Verlet integrator approximates $d^2\mathbf{W}_i/dt^2$ as the simple central difference $0.5\,[\mathbf{W}_i(t + \Delta t) - \mathbf{W}_i(t - \Delta t)]\,/\Delta t$ and the related error is of the order of $(\Delta t)^4$. The Verlet integrator provides good numerical stability, as well as other properties that are important in physical systems such as time-reversibility and preservation of the symplectic form on phase space, at no significant additional computational cost. The study of these systems (including real gases and condensed systems) is usually performed by defining several cell-confined subsystems (a cell is typically a cube in 3D) containing a fixed number of bodies (say N) whose classical mechanics evolution is followed until one of them leaves the cell. At this point, the constance of N is enforced by considering that each cell is surrounded by its replicas and that any outgoing body is replaced by its clone incoming from the corresponding position of one of the surrounding cells.

Important for a proper molecular dynamics study is the correct definition of the initial conditions, of the nature of the ensemble and of the collective indicators of the ensemble of bodies considered. As already mentioned, this is particularly simple for bimolecular collisions of simple molecules for which quantum-like states of the (vibrational and rotational energies) reactants and products can be associated with atom positions and momenta. However, this is not so for very large ensembles of interacting bodies and large molecules. In this case, in order to calculate the average value $\langle P \rangle$ of the observable property P rather than using the corresponding quantum expression

$$\langle P \rangle = \frac{\sum_i e^{-E_i/k_B T} \langle i \,|P|\, i \rangle}{\sum_i e^{-E_i/k_B T}} \tag{5.19}$$

one should adopt its classical statistical mechanics equivalent for which the property to be calculated is formulated as

$$\langle P \rangle = \int P(\{\mathbf{W}\})F(\{\mathbf{W}\})\mathrm{d}\{\mathbf{W}\} \tag{5.20}$$

where

$$F(\{\mathbf{W}\}) = \frac{exp\,[-H(\{\mathbf{W}\})/k_B T]}{\int exp\,[-H(\{\mathbf{W}\})/k_B T]\,\mathrm{d}\{\mathbf{W}\}} \tag{5.21}$$

is the Boltzmann distribution, $U(\{\mathbf{W}\})$ the internal energy and $\{\mathbf{W}\}$ the $\mathbf{r}^N$ (coordinates) and $\mathbf{p}^N$ (momenta) of the N particles system. The value of these integrals are usually estimated using the Monte Carlo technique (i.e., by sampling randomly the various configurations of the system) possibly associated with an importance sampling factor (i.e., by giving a weight to each sampled point). The Monte Carlo technique has been already considered for electronic structures at the beginning of Chap. 3 where it was pointed out that if one needs only the ratio of two integrals (of which one is referred to a given reference configuration like in the case of a transition between two states). the Metropolis method can be adopted. In the Metropolis method a random walk is constructed moving out of the region of space associated with a given configuration (where the integrand is nonnegligible). In the random walk we introduce subsequent random displacements according to some ad hoc criteria (like the one requiring that at equilibrium the number of accepted moves from a state to any other state is exactly canceled by the reverse move).

5.2.2 *Some Popular Molecular Dynamics Codes*

As already mentioned the clear advantage of classical mechanics techniques is their easy extensibility to large numbers of atoms (of the order of several millions or more).

This has fostered the assemblage and development of several popular computer codes designed for classical (and SC) mechanics studies.

Among the most popular codes of this type are:

VENUS96 [110] a program in continuous development by W.L. Hase of the Technical University of Texas that calculates classical trajectories and resulting detailed probabilities, cross section and rate coefficients of colliding bodies (atoms and/or molecules) by integrating related Hamilton equations in cartesian coordinates. VENUS96 discretizes the probabilities using approximate means and can be easily linked to programs performing an SC-IVR evaluation of state-to-state probabilities.

SC-IVR [111] a semiclassical initial value representation program based on the outcome of a classical trajectory package used to calculate the discrete spectrum of medium size molecules and other state-to-state transition properties.

DL_POLY [112] a popular code for the integration of the classical equations of motion of molecular dynamics. It is a general purpose package of subroutines, programs, and data designed to facilitate molecular dynamics simulations. DL_POLY is continually developed at Daresbury Laboratory by W. Smith and I.T. Todorov under the auspices of the British EPSRC and NERC in support of the CCP5 program. It can be used to simulate a wide variety of molecular systems including simple liquids, ionic liquids and solids, small polar and nonpolar molecular systems, bio- and synthetic polymers, ionic polymers and glasses solutions, simple metals, and alloys.

GROMACS [113] is a versatile molecular dynamics package integrating the Newton equations of motion for systems with hundreds to millions of particles. It has been used in a large number of case studies and it consists of a complete workflow aimed at exploiting the interoperability within a local cluster platform (HPC capable) and a worldwide distributed computing infrastructure (DCI) as will be described in some detail later in this chapter). In the workflow, the possibility of coupling the run of different jobs is taken care by means of links (semaphores) defining the dependency job chain.

NAMD [114] is a parallel molecular dynamics code designed for high-performance simulations of large biomolecular systems and it has been used to study the behavior of a lipidic bilayer in water. Ported on the distributed environments using OpenMPI parallel libraries, a direct acyclic graph (DAG) has been implemented to run the code in a semiautomatic way and facilitate the user to carry out his/her calculations.

AutoDock [115] is a suite of automated docking tools. It is designed to predict how small molecules, such as substrates or drug candidates, bind to a receptor of a known 3D structure. Current distributions of AutoDock consist of two generations of software: AutoDock 4 and AutoDock Vina. AutoDock 4 actually consists of two main programs: autodock performs the docking of the ligand to a set of grids describing the target protein; autogrid precalculates these grids. In addition to using them for docking, the atomic affinity grids can be visualized. This can help, for example, to guide organic synthetic chemists to design better binders. AutoDock Vina does not require choosing atom types and precalculating grid maps for them. Instead, it rapidly calculates the grids internally, for the types of atoms needed.

CADDSuite [116] is a code that offers modular tools for most commonly used tasks in the field of computer-aided drug design, that all have the same interface and can easily be used to create even complex workflows. There are algorithms and tools for data storage and retrieval, data preparation, chemical checks, QSAR, Docking, Rescoring, analysis of results. CADDSuite has also been integrated into the workflow system Galaxy, in order to make submitting jobs to different environments or creating, modifying and starting workflows for the user. In essence, a user can thus easily create drug design pipelines directly from a web browser, without any need for software installations on his local computer.

FlexX [117] is a code that predicts within a few seconds the geometry of the protein–ligand complex for a protein with known three-dimensional structure and a small ligand molecule. The use of an intuitive GUI permits the set up of docking runs within a single minute and provides you with fast visual feedback. FlexX can screen a library of 1 000 000 compounds in a few hours on a 30-node cluster. The new screen module also allows you to filter out false positives on the fly. If you are screening compounds from a combinatorial library, you can take advantage of a novel pharmacophore-based combinatorial docking to further gain speed-up and enrichment.

DESMOND [118] is a computer program that can compute energies and forces for the standard fixed-charged force fields used in biomolecular simulations. A variety of integrators and support for various ensembles have been implemented in the code, including methods for thermostatting (Andersen, Nose-Hoover, and Langevin) and barostatting (Berendsen, Martyna-Tobias-Klein, and Langevin). Ensembles typically used in membrane simulations (constant surface area and surface tension) and semi-isotropic and fully anisotropic pressure coupling schemes are also available. Desmond supports algorithms typically used to perform fast and accurate molecular dynamics. Long-range electrostatic energy and forces are calculated using particle-mesh-based Ewald techniques. Constraints, which are enforced using a variant of the SHAKE algorithm, allow the time step to be increased. These approaches can be used in combination with time-scale splitting (RESPA-based) integration schemes. The Desmond software includes tools for minimization and energy analysis (which can be run efficiently in a parallel environment), methods for restraining atomic positions as well as molecular configurations, support for generating a variety of periodic cell configurations, and facilities for creating accurate checkpoints and restart.

AMBER [119] is a code that can compute potential energies for molecular systems (with particular focus on biosystems) by formulating the related potential energy as a combination of terms representing covalent bonds (depending on internuclear distances), bending distortions (depending on planar angles), dihedral and torsional motions (depending on spatial angles), dispersion effects (depending on Van der Waals interactions) and electrostatic forces (depending on charges and distances). The software comes with libraries and databases of parameter values tailored to suit different molecular systems.

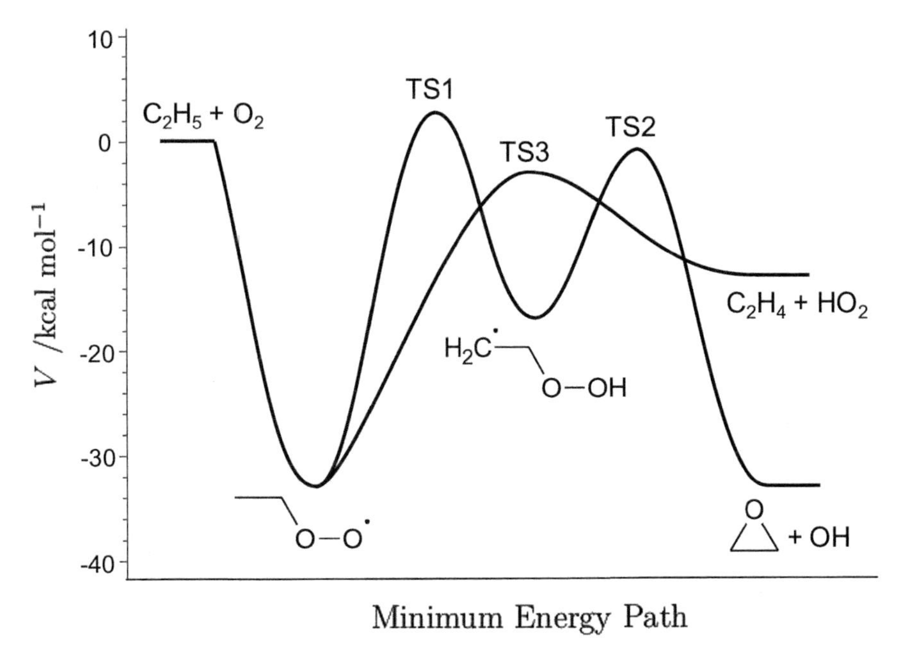

Fig. 5.5 The MEPs of the $C_2H_5 + O_2$ reaction corresponding to the product $C_2H_4 + HO_2$ and $C_2H_4O + OH$

5.2.3 *Force Fields*

Other approximations are usually adopted in molecular dynamics treatments of large systems in addition to the already mentioned dropping of discretization of bound motions and of the uncertainty constraints on the values of the conjugated variables associated with the use of classical mechanics. The most limiting one is associated with the difficulty of accurately handling the molecular interactions of large and complex systems. About this, we have already pointed out for three-body systems that microscopic branchings (i.e., the exploration of separate regions of the PES) by trajectories originating from different initial conditions (like attacks from different sides of the molecule) prompt higher level of calculations to better define the potential energy channels of the PESs.

For illustrative purposes, we show in Fig. 5.5 the case of the nine atoms elementary reaction $C_2H_5 + O_2$. In the figure, the two main (low energy) MEPs are plotted to the end of showing the clear macroscopic branching between the H abstraction (leading to the $C_2H_4 + HO_2$ products) and the O_2 insertion into the double bond (leading to the $C_2H_4O + OH$ products). Moreover, the abstraction MEP shows a double barrier sandwiching a fairly deep well. Inevitably, the accurate determination of these electronic structure features requires extended ab initio calculations on a fine multidimensional grid of molecular geometries. Obviously, for larger or more variegated molecules the MEPs may become more structured and richer of alternative paths making the definition of the functional representation of the overall PES almost impossible.

Accordingly, thanks to the present easier accessibility to fast computers, increasing use is made of "on the fly" techniques (also called direct) in which the potential energy of a given molecular geometry is computed using a suitable package right when (and if) it is actually needed rather than adopting a general functional formulation of the whole PES. This approach avoids heavy ab initio calculations of the potential energy (and related derivatives) for the system geometries which are not reached during the integration of the dynamical equations. The price to pay, however, is the impossibility of carrying out a preliminary analysis of the PES to discard nonconverged values and the practical impossibility of using top-level theoretical treatments for ab initio calculations and LS fitting.

For complex systems, the most frequently adopted solution is the molecular mechanics (MM) one. Typically, in MM applications an ex ante overall assemblage and calibration of the PES is performed by making use of several simple empirical local formulations of the interaction to shape the energy channels associated with the different degrees of freedom. In this approach, based on the separate treatment of independent simple components (force fields), each bond length, each planar or dihedral or out-of-plane angle, each ionic or dispersion interaction is treated individually despite the fact that this is likely to introduce an uncontrolled fine structure in the interaction representation. The theoretical ground for this approach is the adoption of a generalized MBE of the interaction (as already discussed in Chap. 3) followed by the dropping of terms of higher order. This means, for example, that the retained components of molecular motion (see Fig. 5.6) are:

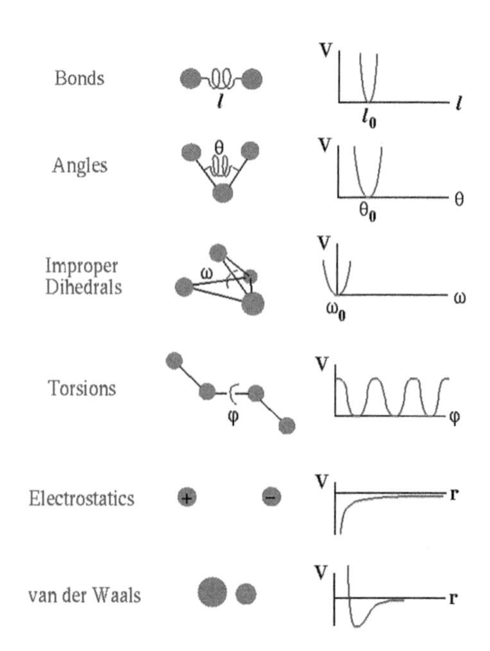

Fig. 5.6 The most common force-field terms

- the two-body stretching of atom–atom bonds,
- the bending of an in-plane angle formed by two bonds,
- the variation of either a proper or improper dihedral angle,
- the torsion around a given axis
- the Coulombic repulsion/attraction,
- the short-range repulsion and long-range attraction of two dressed nuclei.

which are formulated, respectively, using simple analytical functions like:

- either a Harmonic or a Morse oscillator,
- a Harmonic oscillator,
- an out-of-plane Harmonic oscillator,
- a Fourier series,
- a positive/negative inverse r power, and
- an inverted 12th power of r (or a Buckingham Ae^{-Br}) at short range and a multipolar expansion of the van der Waals type at long range

Further components can be H-bond or other terms (including crossed ones) useful to formulate particular interactions. Accordingly, the usual formulation of the overall interaction of N bodies in the standard AMBER approach using the annotation of Fig. 5.6 reads as

$$V(r^N) = \sum_{bonds} \frac{1}{2} k_b (l - l_0)^2 + \sum_{angles} \frac{1}{2} k_a (\theta - \theta_o)^2$$
$$+ \sum_{dihedrals} \frac{V_n}{2} [1 + \cos(n\omega - \gamma)] + \sum_{j=1}^{N-1} \sum_{i=j+1}^{N} \left\{ \varepsilon_{ij} \left[x_{ij}^{12} - 2x_{ij}^{6} \right] + \frac{q_i q_j}{4\pi\varepsilon_0 l_{ij}} \right\} \tag{5.22}$$

with $x = l_0/l$. In some cases additional terms like

$$V_{H-bond} = \sum_{H-bonds} \left[\frac{C_{ij}}{R_{Hij}^{12}} - \frac{D_{ij}}{R_{Hij}^{10}} \right] \tag{5.23}$$

and

$$V_{\phi} = \frac{1}{2} \sum_{out-of-plane-bends} K_{\phi} \phi^2 \tag{5.24}$$

are introduced in order to include H-bonds (R_H is the related distance) and out-of-plane bends. In other approaches different decompositions of the already discussed V_{inter} and V_{intra} molecular interaction terms are used.

Obviously, the parameters of the formulation, whenever possible, are optimized to reproduce the experimental data (scattering, spectroscopy, etc.) and/or ab initio calculations. Yet, one of the main problems of this approach is the fact that the reference force-field geometries are difficult to change. Sometimes, for large molecules, one may not even want to derive the complete detailed information on initial and/or final states. For example, in the case in which one wants to know the more likely transient or transition state the efficiency of a reaction at a given temperature T, the dimensionality of the problem can be simplified by compacting the description of one or more clusters of atoms not directly involved in the process into a single body. Obviously, when more accuracy is needed, mixed quantum (QM) and molecular dynamics (MD) methods can be used.

5.2.4 *Toward Multiscale Treatments*

As already mentioned, elementary processes are often to be considered as building blocks of more complex (multiscale) procedures in which numerical estimates obtained from a (reasonably accurate and realistic) description of a scientific and/or technological formulation of the problem are accompanied by additional treatments. These additional treatments are often concerned with the reduction of the molecular granularity thanks to the clustering of more atoms in a single body. This requires an organization of the related computational procedures in workflows structured both horizontally (for computations occurring at the same level of granularity) and vertically (for computations occurring at different levels of granularity).

In chapter one we have already considered, in this respect, the disentangling of computational complexity in combustion processes and the disentangling of complexity arising from higher scale kinetic treatments coupling several elementary processes. Here, although this is not a goal of the present book, in order to give a detailed account of multiscale methods, we refer to another technological application in which the detailed microscopic (atomistic) level considered for the elementary processes of small molecules is mitigated by the application of higher scale statistical treatments. This is the case, for the example we consider here, of studies of the properties of several gaseous systems for which use is made of direct simulation Monte Carlo (DSMC) techniques [120]. DSMC leverages on probabilistic (Monte Carlo) simulations to solve the Boltzmann equation for finite Knudsen number (Kn) fluid flows. The method is of widespread use in the modeling of rarefied gas flows in which the mean free path of a body is of the same order (or greater) than a representative physical length scale often expressed in terms of Kn that is given by a dimensionless number defined as the ratio between the molecular mean free path length λ and a physical length (L) like the radius of the bodies forming the fluid. For example, for a Boltzmann gas, the mean free path is given by

$$\mathrm{Kn} = \frac{k_B T}{\sqrt{2}\pi d^2 p L} \tag{5.25}$$

with k_B being the Boltzmann constant, T the thermodynamic temperature, d the particle hard sphere diameter, p the total pressure. For particle dynamics in the atmosphere, when assuming standard temperature and pressure (i.e., 25 °C and 1 atm) one has $\lambda \equiv 8 \cdot 10^8$ m. In supersonic and hypersonic flows rarefaction is characterized by the Tsein's parameter, that is equivalent to the product of the Knudsen and the Mach (M) number (KnM or M^2/Re) with Re being the Reynolds number.

The DSMC method models the flow of a fluid in terms of colliding bodies made by a large number of molecules and solving the related Boltzmann equation. Bodies are moved through a simulation of physical space in a realistic manner that is directly coupled to physical time. Interbody and body-surface collisions can be calculated using probabilistic, phenomenological and collisional models. The method finds application to several technologies including the estimation of the Space shuttle reentry aerodynamics and the modeling of microelectronic-mechanical systems using appropriate interfaces (see Refs. [121–124]).

5.3 Supercomputing and Distributed Computing Infrastructures

5.3.1 High-Performance Versus High-Throughput Computing

In general, computational chemistry has grown by exploiting at any time the best performing compute machines available. As already mentioned, the real rise of computational chemistry to the dignity of separate discipline, has occurred only with the advent of the so-called mainframes (one (CPU) to many (users) compute platforms) based on ad hoc designed advanced CPUs and characterized in the field of molecular sciences for the ability of carrying out high-performance off-line FORTRAN number crunching dominant calculations. This has occurred in the second part of the 20th century when the mainframes were offering increasing compute power on single CPU architectures by continuously improving circuitry efficiency (shorter clock-period, faster electronics, larger buses and higher miniaturization), components quality (communication bandwidth, size and speed of caches, size and efficiency of memories), optimized use of the processor (multiprogramming, time sharing, look ahead and prefetching) often driven by progress in science research. As just mentioned, the reference language was FORTRAN that evolved through different versions to the highly popular FORTRAN IV (1961), FORTRAN 66, and FORTRAN 77. More recent versions are those of 1990, 1995, 2003, 2008, and 2015 increasingly tailored to suit the HPC needs.

Further speed-enhancing progress was associated with architectural changes to the organization of different levels of cache memories and with the design of innovative languages and operating systems. The most significant advances were associated with concurrent execution at both operation and instruction level, multiple functional units and pipelines, super- and multiscalar microarchitectures, long instruction words, etc. At the same time, the exploitation of architectural innovation at such micro level significantly enhanced the possibility of faster parallel runs of highly coupled procedures pushing computer architectures into the era of present supercomputers. In molecular sciences this has meant in particular the enhancement of the performances of electronic structure calculations based on matrix manipulations and linear algebra operations despite their mixed heterogenous nature born out of subsequent stratifications of different software components. After all, as already pointed out, this has happened during times in which scientific interests (including those in molecular science) were the driving force for advancing the design of computer architectures.

The decisive leap forward, however, in molecular science computing has leveraged on the introduction of multicore and multi-CPU machines prompted by more market-oriented consumer electronics.

Important steps of this process were the assemblage of single instruction stream multiple data stream (SIMD) platforms (array processors, vector computers, etc.) in which, as shown in Fig. 5.7, a single control unit (CU) instructs the different processing units (PU)s to perform the same operation on different datasets (DS)s and store results in memory (MM) coupled with their replication within the same processing unit as well as Multiple Instruction stream Multiple Data stream (MIMD) platforms (the truly parallel machines) in which, as shown in Fig. 5.8, there is a CU for each PU disentangling the operations from the constraint of being the same for all the PUs. Several variants of these platforms are commercially available depending on whether they have shared or local memory (either physically or logically implemented) and dedicated or shared (multilevel, i.e., on chip, on card, on blade, on tower, and on clusters of towers) networking. The management of multiple processors has made also significant progress by resorting to specific software tools like High-Performance FORTRAN, Parallel Virtual Machine (PVM), and Message Passing Interfaces (MPI). A picture of three types of machines having marked the evolution of concurrent computing is given in Fig. 5.9.

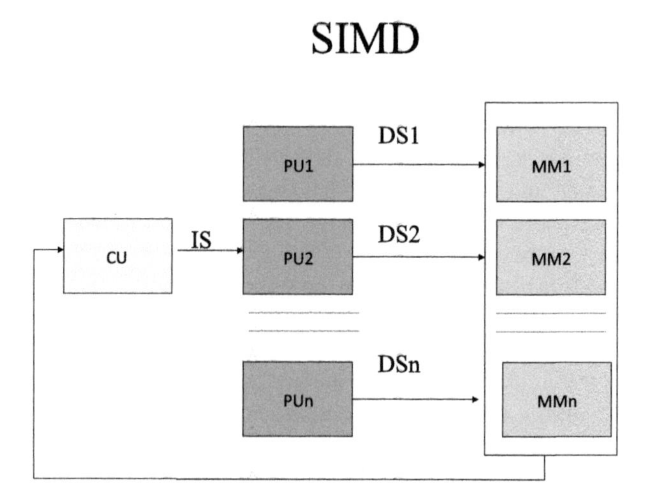

Fig. 5.7 The SIMD scheme

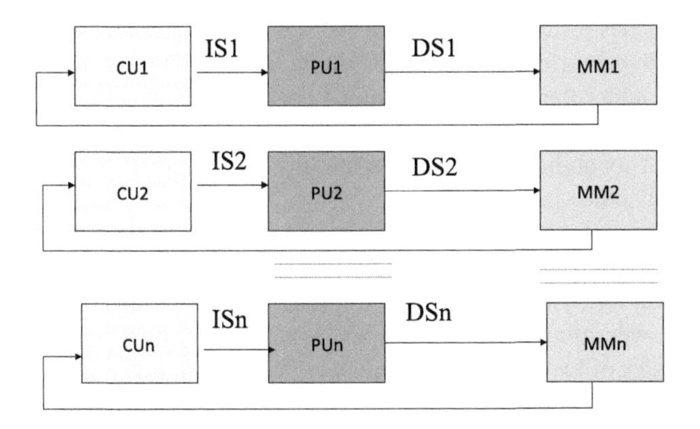

Fig. 5.8 The MIMD scheme

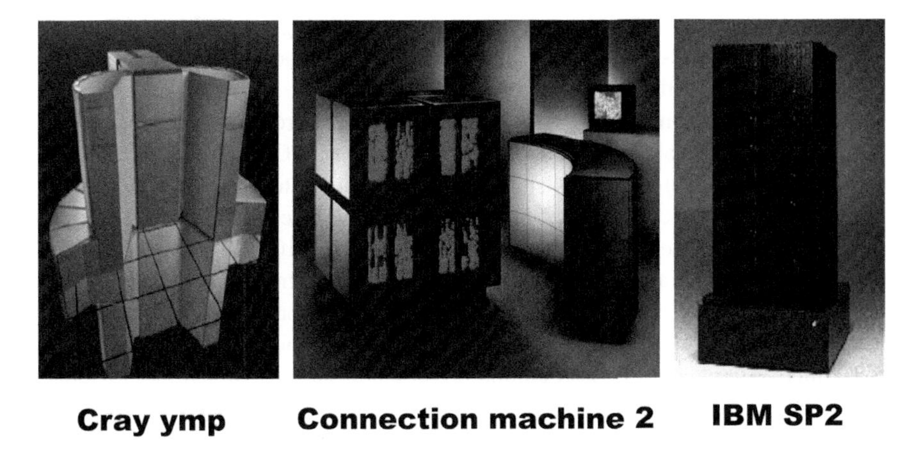

Fig. 5.9 A picture of three most popular parallel supercomputers of the 1990

However, in order to enhance the performances of the already mentioned electronic structure codes on these platforms, an in-depth significant reorganization of the applications is needed. Thanks to the restructuring and/or the afresh design of new codes (see for example Ref. [17]) modern platforms can, indeed, target performances of the order of ExaFLOPS also in molecular science applications.

5.3.2 Networked Computing and Virtual Communities

A different evolution line of computing that has impacted significantly molecular science studies is that of the proliferation of minicomputers and desk computers (personal computers (PC) and workstations) as local networked resources having the clear advantage of narrowing the distance between compute resources and users. First of all, this has resulted into a specialization of local resources for user-specific needs thanks to user-oriented operating systems. Among these, Linux (born in 1991 and operating under the GPL licence based on the software of the GNU project). Linux became very popular not only because of its ability to emulate terminals but also because of its ability to read and write files from/to disks and to act as a true kernel able to handle operating systems.

Further progress was made by enabling Linux to execute the X server and to provide an integrated system for graphic interfaces. At present Linux has become the preferred operating system for servers in production environments and embedded devices. Moreover, Linux has also a strong presence in the market of scientific desktops. As a matter of fact, Linux is the most popular operating system for executing Apache, MySQL and PHP, the software grounding most of the web servers worldwide and has developed as well desktop environments interfaces similar to those of Microsoft Windows and Mac OS X closer to the needs of the users.

The combination of PC-like user friendliness with high network connectivity has enhanced the possibility of clustering remote local platforms and fostered the development of high-throughput computing (HTC).[2]

At the end of the mentioned three subsequent EGEE projects, a European HTC powerful distributed platform (named EGI, the European Grid Infrastructure managed by EGI.eu [125]) was established in order to coordinate a large number of geographically dispersed compute resources connected over the public network through the use of appropriate middleware and tools (see Fig. 5.10). As a result, GC has become an important asset of the European scientific community enabling the concurrent execution (over hundreds of thousands of processors) of several distributed programs for applications made of decoupled or loosely coupled tasks.

An innovative feature of grid computing is the possibility of setting new goals to scientific research and technological applications by aggregating a large number of highly dispersed and heterogeneous small size computers to the HPC ones (like PRACE for the EU and XSEDE for the US) made of several millions of cores and large storages. The main problem in this is the distance between the policies adopted by the management of large-scale compute facilities and the expectation of a large fraction of the Molecular science users (especially those having a high activity of design and development of innovative codes). The compute time allotment policy

[2]High-throughput computing (HTC) refers to machines exhibiting an efficient execution of a large number of loosely-coupled tasks. HTC systems are independent sequential jobs that can be individually scheduled on many different computing resources across multiple administrative boundaries. HTC systems achieve this using various grid computing (GC) technologies and techniques. In Europe, a strong impulse to HTC has been given by the last Framework Programmes, especially as a support to the High Energy Physics transnational community initiatives, by funding several international collaborative projects like DATATAG (http://datatag.web.cern.ch/datatag/), EGEE-I-II-III (http://www.egee.eu), WLCG (http://wlcg.web.cern.ch/), EGI-Inspire (https://www.egi.eu/about/egi-inspire/), and EGI-Engage (https://www.egi.eu/about/egi-engage/).

Fig. 5.10 A sketch of the geographical location of the most important nodes of the European grid

adopted by the mentioned large-scale facility networks, in fact, privileges compute time allocation plannings rewarding massive (systematic) production runs selected through individual project-based competitions maximizing a constant use of the machines. The prevalent view of the community members prefers, instead, an a la carte use of the machines that combines design and development of innovative codes through a collaborative exploitation and reuse of the expertise and achievements of different researchers.

The established structures of the scientific communities are the so called Virtual Organizations (VO)s[3] and the Virtual Research Communities (VRC) [126].[4] Both VOs and VRCs nurture the evolution of scientific computing and sustain the progress of the users' scientific activities by establishing and supporting Virtual Research Environments (VRE)s [127].[5] This approach encourages thematic communities to develop elementary and composite workflows of codes of different nature as well as to design collaborative metaworkflows to meet the increasing demand of higher complexity accurate simulations.

[3]VOs are groups of researchers bearing similar scientific interests and requirements being able to work collaboratively with other members. VO members share resources (e.g., data, software, expertise, CPU, and storage space) regardless of their geographical location and join a VO to the end of using the grid computing resources provided by the resource provider. According to the VO's requirements and goals, EGI (European Grid Infrastructure) [125] provides authentication, job allocation, activities monitoring support, services and tools allowing them to make the most of their resources.

[4]VRCs are self-organized research communities which give individuals within their community a clear mandate to represent the interests of their research field within the EGI ecosystem. They can include one or more VOs and act as the main communication channel between the researchers they represent and EGI.

[5]VREs are e-infrastructures used for e-science operations that maximize coordination and identify commonalities across the European research infrastructures as well as common solutions to problems, which can then be implemented by the involved communities thanks to synergies across member states, while minimizing overlap of effort. VREs can automate aspects of data recording, including metadata, using customisable workflows.

5.3.3 *The Collaborative Grid Empowered Molecular Simulator*

Since the first debut for Physics-based sciences and technologies, GC has been gradually extended to other disciplines leveraging on the formation of the already mentioned VOs (among which the COMPCHEM VO of the molecular science community with interests in theoretical and experimental studies on chemical reactions (see Ref. [128]) within the EGEE project). An important feature of the VOs is that they set by themselves their targets and manage their own membership according to the internal requirements and goals. EGI provides support, services, and tools to allow VOs to make the most out of the available resources. EGI hosts more than 200 VOs for communities with interests as diverse as earth sciences, computer sciences and mathematics, fusion, life sciences, or high energy physics. Moreover, members of VOs can access, among the already implemented applications, those of interest for their work and can port on the shared environment applications and tools for personal and/or community use. In other words, forming a VO and using the related DCI, leads naturally to the sharing of HW and SW and to the development of cooperative mechanisms (https://www.egi.eu/community/vos/) as it has happened for COMPCHEM.

As already seen, more general organizations are the VRCs which bear the clear mandate to represent the interests of a research community within the EGI ecosystem. They can include one or more VOs and act as the main communication channel between the researchers they represent and EGI. EGI establishes partnerships with individual VRCs through a memorandum of understanding (MoU). Following the accreditation process and final agreement, VRCs can access the computing resources and data storage provided by the EGI community through open-source software solutions. VRC members can store, process and index large datasets. They can also interact with partners using the secured services of the production infrastructure of EGI. The first set of VRCs to have signed a MoU or a letter of intent (LoI) with EGI are: WeNMR (structural biology), LSCG (life science), HMRC (hydrometeorology), LHC (high energy physics), CLARIN, and DARIAH (arts and humanities) (see https://www.egi.eu/community/vos/vrcs/). Next EGI has negotiated agreements with other research communities including the Chemistry, Molecular and Materials Sciences and Technologies (CMMST) one and this has led to the establishment of the homonymous VRC. The mission of the CMMST VRC, like that of other thematic communities, is to build a specific VRE aimed at orchestrating the activities of both the e-infrastructure experts and the molecular and materials researchers so as to enable an effective intra- and trans-community networked implementation and coordination of a collaborative/competitive environment. The collaborative/competitive environment is specifically designed to allow both:

- a discovery of the compute resources and a selection based on quality parameters and automated access,
- the accessibility to software libraries and their coordinated usage,
- the use of specialized web portals and the reuse and production of data and know how, and

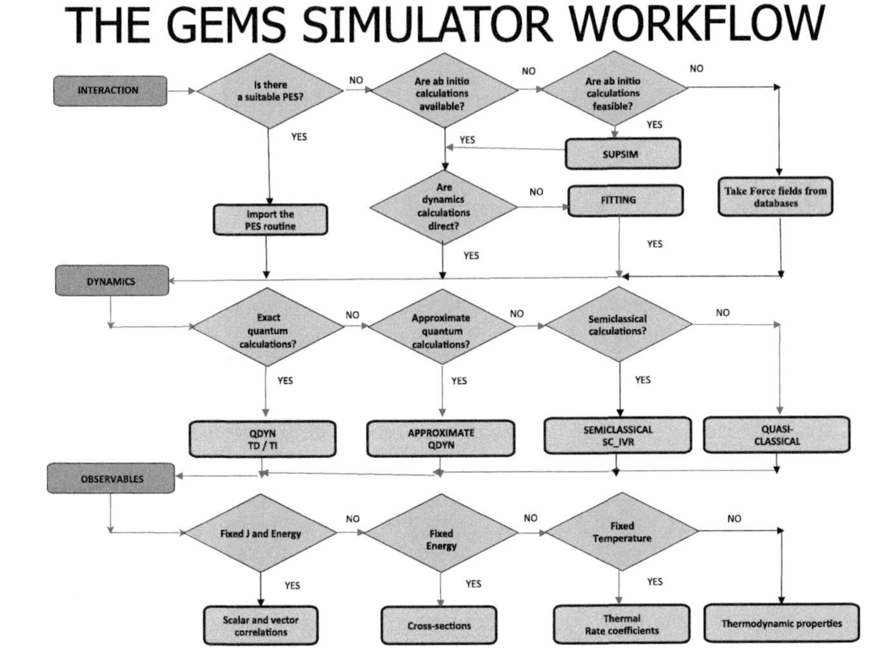

Fig. 5.11 A sketch of the finer level articulations of GEMS in specific packages

• the monitoring of collaborative activities and the rewarding for the work done on behalf of the community.

A typical service tailored to the needs of the molecular science community for the simulation of the observables of chemical processes is the generalization of GEMS [107] as a metaworkflow. As shown in Fig. 5.11, GEMS manages, in fact, state-of-the-art ab initio electronic structure and nuclei dynamics compute programs aimed at calculating the basic quantities needed to accurately simulate experimental measurables of light-matter and matter-matter apparatuses. This level of service, central to the activities of the CMMST community, provides, as already seen for the "last mile" simulation of crossed beam experiments [108], ab initio information on molecular geometries and energies (INTERACTION module) fitted PESs (FITTING module), dynamical properties (DYNAMICS module) and measured quantities (OBSERVABLES module) each articulated in different options. Indeed, the concerted usage of highly accurate electronic structure and nuclei dynamics calculations is based on the collaborative use of different compute resources and technologies and on the agreed definition of (at least de facto) data format standards-based like those developed within some COST projects. They represent the most advanced research ground for both methodological ab initio developments (including the design and testing of new concurrent algorithms) and the rationalization of molecular structures and processes.

5.4 Toward an Open Molecular Science

5.4.1 A Research Infrastructure for Open Molecular Science

Although chemistry is one of the oldest research European communities and its research discoveries have a high impact on the evolution of other disciplines like biology, medicine, materials, aerothermodynamics, etc., it has not a significant representation within the organized GC, grid computing, and DCI, distributed compute infrastructure, initiatives and no share of the Horizon 2020 funding for projects and research infrastructures (RI). For this reason, it is important to discuss in this last section of the book which of the strategic lines of the present evolution of compute technologies offers a future to chemical process studies. In particular we shall discuss here the roadmap for establishing an open molecular science RI by analyzing in details the contents of SUMO CHEM the most recent proposal submitted for European funding by the molecular science community.[6] As apparent from both the text of the proposal and the COMPCHEM VO and CMMST VRC reports within the activities already carried out during the EGEE and EGI-Inspire projects, a central target of the molecular science community is the establishment of a European RI pioneering a new way of collaborating between computational and experimental chemists by creating a seamless open environment for joint research, data production and reuse including its transfer in innovation and societal utilization. Such openness is built on the understanding of methods and applications relevant to the activities of the members of the molecular science community. The molecular science community uses data in order to exploit the considerable potential of available research facilities and e-infrastructure to jointly run sophisticated experiments and simulations target-

[6]SUMO-CHEM: "Supporting Research in computational and experimental chemistry via Research Infrastructure" submitted to the Horizon 2020 framework call H2020-INFRAIA-2016-2017 (Integrating and opening research infrastructures of European interest) by Gabor Terstyanszky. Topic: INFRAIA-02-2017 Type of action: RIA (Research and Innovation action,) Proposal number: 731010-1. Published on the NEWS issue of the e-magazine VIRT&L-COMM of Sept. 2016 http://www.hpc.unipg.it/ojs/index.php/virtlcomm/issue/view/17. As from its abstract of the submitted proposal, SUMO-CHEM is an open molecular science initiative that "will integrate research facilities and infrastructures with computing and data resources into the SUMO-CHEM RI to enable joint research involving computational and experimental chemistry and other research communities. This RI will have an open architecture to allow its extension with further research facilities and resources to be used by the chemistry and other communities. The SUMO-CHEM RI will allow researchers and developers to run industrial simulations and scientific experiments using European, regional and national research facilities and e-infrastructure resources through an intuitive and seamless virtual access considering different levels of their expertise and skills. The major innovation of the project will be in management of scientific data covering the whole life cycle of data using metadata, ontologies, and provenance based on advanced data and computing services. SUMO-CHEM will enable and support multidisciplinary research in cooperation with ESFRI and other major research initiatives to address climate and energy societal challenges."

Although positively evaluated for the idea of connecting experimental and computational chemistry communities and infrastructures, for its capacity to go beyond the state of the art for the proper selection of use-cases and for its multidisciplinarity and development of networking activities, the proposal was not funded.

ing societal of the society. This will help pooling the intellectual efforts in creating and refining data management approaches, such as data preservation, identification, and citation that can be used across various disciplines by surmounting the difficulties associated with the heterogeneity of the different components of the chemistry community.

Members of the molecular science community make use, in fact, of both different types of research facilities to run experiments and different types of e-infrastructure resources to run simulations. They also produce different types of data in different data formats. This heterogeneity has prompted the provision of a user interface that seamlessly hides differences in data, e-infrastructure resources, and research facilities. This implies the design of a *service orchestrator* to manage a set of ad hoc microservices and provide maintainable and sustainable services to assemble a completely flexible and agile approach to front-end development. Researchers will communicate through a dashboard, a web-enabled interactive front-end in the community layer social media like user interface that will offer a new means of reusing, understanding, and further developing for scientific progress either experiments or simulations. In the spirit of openness the dashboard will allow interaction among researchers and simple users allowing a sharing of data at one side and of research facilities and e-infrastructure resources at the other side. Moreover, the Data Service will handle the whole data life cycle including creating, publishing, sharing, curating and preserving data. This will tightly bind research facilities used in experiments and computing and data resources used to run simulations enabling so far a two-way research cooperation between experimental and computational chemistry through publishing and sharing within an open science approach. In other words, researchers can run simulations to verify experimental results and design further simulations based on these results while analyzing simulation results researchers can plan more efficient experiments. Such two-way data exchange to follow new research challenges to create and validate improved or new materials that business, industry, and society can use is, indeed, the real strength of an open science approach that will enrich the chemistry community with new competences and technological solutions addressed to societal challenges.

5.4.2 Foundations and Stakeholders for the Molecular Open Science RI

As already mentioned the proposed open molecular science initiative to establish a European RI are grounded on the good practices implemented by COMPCHEM VO and the CMMST VRC while developing GEMS. They are also supported by some computer science laboratories. Because of its positive impact on the community, of the Cross-disciplinary fertilization with other communities, and of the proposed standardization procedures enabling better offer of services to the wide community and transparency of research data, the initiative of establishing an open molecular science RI sees also the involvement of the European Chemistry Thematic Network (ECTN), of the EUCHEMS (Association for Chemical and Molecular Sciences) through its Computational Chemistry division, of the European Joint Doctorate in Theoretical Chemistry Modelling and Computational (TCCM) ITN-EJD-642294 in particular for the generic community members activities and for training and educational aspects as well.

Accordingly, the main target of the project is a significant evolution of the collaborative use of molecular science computational and experimental techniques enabling an accurate determination of intra- and intermolecular interactions and the development of rationales for driving molecular processes to produce innovation. The latter objective leverages on the measurement and calculation of detailed structural and dynamical properties of elementary chemical processes occurring in gas and in condensed phase. It provides the knowledge necessary for grounding further studies on the evolution in time of interleaved elementary processes in kinetics, as well as their combination with statistical, fluid dynamics and/or condensed phase treatments allowing the accurate modeling of important phenomena and to the development of innovative technological solutions to important societal challenges, such as climate change, green energy, food security.

Direct support to the project is given by the following large European laboratories:

- ELETTRA-SINCROTRONE Trieste specialized in generating synchrotron (Elettra) and free-electron laser (FERMI) radiation enabling the characterization of material properties and functions;
- LENS specialized in providing short-pulse laser sources as experimental facilities for spectroscopic and nonlinear optics research;
- FLASH Free-Electron LASer set for VUV and soft X-ray radiation operated in the “self-amplified spontaneous emission” (SASE) mode;
- PETRA accelerator based on an X-ray source to run pump-probe X-ray absorption experiments with high photon flux;
- BEAMLAB coordinated set of crossed molecular beam and beam gas apparatuses laboratory allowing generation and velocity selection of reactant beams;
- BEYOND-NANO laser-induced plasma and phenomena under nonequilibrium conditions reactor;
- CLUR cluster of high power pulsed lasers in combination with multiphoton ionization laser spectroscopy and time-of-flight mass spectrometry; and
- CNRS and CNR advanced French and Italian research facilities on plasmas, combustion, and renewable energies exploitation.

Specific activities for the dissemination and exploitation of produced results are carried out jointly with the already mentioned thematic associations and associated SMEs. In particular, the transnational DRAG [129] cluster is supporting e-learning activities and Master-UP (http://www.master-up.it/) is supporting innovation technology transfer [130] (see also later in this section) with the commitment of achieving societal impact and sustainability. Moreover, the requirements issued by the project partners are regularly collected and analyzed in order to be adequately prioritized to the end of maximizing the interoperability with the existing e-infrastructures.

5.4.3 Compute Resources and Data Management for Molecular Open Science

Computational activities of the molecular science community rely on the usage of the EGI Federated Cloud and PRACE resources. These compute resources are complemented by those of the distributed European Infrastructures and vocational national and regional facilities (ranging from multicore processors to cloud clusters) among which the most important are as follows:

CMAST a Virtual Laboratory to support research in chemistry integrated into the computing infrastructure CRESCO (a production grid of computational resources belonging to ENEA DTE-ICT).

RECAS a consortial of four national Italian data centers (Napoli, Bari, Catania, Cosenza) that is part of the European Grid Infrastructure (EGI) and INFN.

MOSGRID a compute infrastructure that provides Grid services for molecular simulations leveraging on an extensive use of the German D-Grid-Infrastructure for high-performance computing to handle metadata and provide data mining and knowledge generation.

Leveraging on these resources and on existing tools (see also the next subsection) an Open Science Data Cloud (OSDC) can be established for the molecular science community. The OSDC allows scientists to manage, analyze, share, and archive their datasets. Datasets can be downloaded from the OSDC by anyone. The OSDC is not only designed to provide a long-term persistent home for scientific data, but also to provide a platform for data-intensive science so that new types of data-intensive algorithms can be developed, tested, and used over large amounts of heterogeneous scientific data. The not-for-profit OSDC is articulated as follows:

(1) Use a community of users and data curators to identify data to add to OSDC.

(2) Use permanent IDs to identify this data and associate metadata with these IDs.

(3) Support permissions so that colleagues can access this data prior to its public release and to support analysis of access-controlled data.

(4) Support both file-based descriptors and APIs to access the data.

(5) Make available computing images via infrastructure as a service that contains the software tools and applications commonly used by a community.

(6) Provide mechanisms to both import and export data and the associated computing environment so that researchers can easily move their computing infrastructures between science clouds.

(7) Identify a sustainable level of investment in computing infrastructure and operations and invest this amount each year.

(8) Provide general support for a limited number of applications.

(9) Encourage OSDC users and community of users to develop and support their own tools and applications.

The Open Science Data Cloud (OSDC) currently serves multiple disciplines that use big data, including the earth sciences, biological sciences, social sciences, and digital humanities.

The goal of SUMO-CHEM is to ground the proposed molecular ODSC on a collaborative endeavor between Theorists and Experimentalists in order to ensure not only the exploitation of molecular science but also a significant leap forward in its foundations and ability to interpret natural phenomena. In the molecular ODSC curation, preservation and access to the data will be arranged for its whole life cycle (from creating, publishing, accessing, curating and, preserving) using metadata, ontologies, and provenance. Data resources to be used in experiments and simulations will include European, national and regional ones. The data range from experimental measurements recorded in the own lab, over large infrastructure facilities to specific analysis and simulation applications. All these resources, usually stored in local structures using nonstandard formats, are proposed for standardization using metadata in order to facilitate the researchers work, to improve interoperability and to enhance resilience. A uniform, open metadata format accompanied by robust ontologies, is necessary both for primary experimental data to annotated simulation outcomes. Metadata annotation employing a standardized markup format and corresponding ontologies is needed to handle the multitude of data and formats. It will be hosted in a distributed infrastructure that will also store the data and make it available persistently to the community through the usage of standard services. In order to improve reproducibility and reusability, the data will be curated by the related use-case that will make it available for the scientific community together with the protocols used for its generation.

5.4.4 Molecular Open Science Use-Cases

The typical sectors of activities and subcommunities of the molecular science community chosen by the SUMO-CHEM open science RI project are as follows:

use-case 1: Chemical dynamics and energetics. This use-case will utilize beamlines of the Synchrotron and Free-Electron Laser light sources to investigate molecular systems interacting with radiation in a wide range of energies, photon field strengths, and temporal regimes. These experiments will allow researchers to investigate specific properties of matter under selected conditions. Such experiments generate a large volume of data, with computational chemistry being indispensable for the analysis. Ab initio simulation software packages such as ADF, DALTON, MCDTH, NWCHEM, VENUS, etc., running on high-performance computing resources will be used.

use-case 2: Functional and structural properties of matter. This use-case will use femtosecond and nanosecond pulsed lasers in combination with pump-probe and laser spectroscopy, time-of-flight mass spectrometry and ion and photoelectron imaging techniques to study the dynamics, stereo-dynamics, and quantum control of molecular processes including molecular photodissociation and photochemistry and bimolecular reactive and inelastic collisions and material science with lasers. Complementary to experiments researchers will run simulations to study the dynamics of elementary molecular processes using electronic structure calculation software (MOLPRO, MOLCAS, and GAUSSIAN).

use-case 3: Plasma in nonequilibrium conditions. Plasma phenomena in nonequilibrium conditions are currently being experimentally and theoretically studied at the Beyond Nano RI to obtain an efficient use of energy in different applications. The modeling team complements the experimental investigation of plasma by revealing details impossible or very difficult to access in the experimental approach. To solve Boltzmann transport equations (BTE), deterministic (state-to-state molecular dynamics) and stochastic methods packages such as DSMC (direct simulation Monte Carlo) and PIC (Particle-in-Cell) will be ported to the RECAS computational infrastructure. The following in-house developed simulation packages: PLASMA-FLU (plasma simulation), PIC, DSMC, and EPDA (elementary processes data aggregator) will ported to the RI.

use-case 4: Spectrum of metal complexes. Experimentalists will record nonlinear and time-resolved spectra of metal complexes using X-ray absorption, flash laser and linear and time-resolved spectroscopy and compare the results with simulated spectra to find the best matching molecular structure. Computational Chemists will explore the phase space running atomistic simulations for computing free energy surfaces. They will analyze simulation data of metal complexes complementing experiments for vibrational and electronic spectroscopic properties in different environments. There are further simulations related to experiments investigating ground and excited electronic states under controlled conditions of temperature and pressure using linear and time-resolved spectroscopy. These simulations will use NWChem, Gaussian, ORCA, Jaguar, MOPAC, DFTB+, and MNDO99.

use-case 5: Renewable energy storage as chemicals. This use-case will leverage on design of complex kinetic systems involving gas and solid state catalyzed processes using efficiency parameters derived from ab initio studies checked against highly detailed measurements of the corresponding elementary gas phase processes obtained from molecular beam–beam and beam gas experiments. The measurements will also utilize a prototype industrial apparatus, built by a consortium of SMEs to use energy from renewable sources to produce methane from CO_2 and store it in forms easy and safe to transport. The complex kinetics simulations will make use of the ZACROS code complemented by the evaluation of the dynamical properties using the following software packages: APH3D (both time dependent and time independent), ABC, RWAVEPR, and VENUS.

use-case 6: Cleaner combustion. This use-case will focus on design of smart energy carriers based on the COST SMARTCATS proposal to increase fuel flexibility and carbon efficiency of energy production and to support distributed energy generation strategies by bringing together numerical and diagnostic tools. The experimental RI ranges from elementary reactors (sodium-cooled fast and plug flow reactor) and to complex systems (engine and cyclonic burners) enhanced by analytical chemistry techniques (GC/MS, HPLC) and advanced optical diagnostics (spectroscopic and laser-induced fluorescence (LIF) measurements). The simulations based on CRECK, Pope, ANOVA (variance analysis), and Tukey or Dunnett modeling software to complement the experiments by validating the experimental results and optimizing the combustion process.

use-case 7: Secure, clean, and efficient energy production: low carbon technologies. This use-case will develop market affordable, cost-effective and resource efficient solutions for the energy system based on low-carbon technologies through the CMAST virtual laboratory by designing new materials at the nanoscale level, combining experimental and numerical results and speeding up the production of specialized nanomaterials for energy applications. Computer modeling technologies will be used to reveal the microscopic origin of macroscopic properties and will be exploited for both increasing the efficiency of devices producing and storing energy and for lowering the quantity of needed raw materials. The use-case will focus on materials for PV, hydrogen and nuclear technologies in order to enhance their chemical properties at the interface.

use-case 8: Optimization of biodiesel production. This use-case will investigate kinetic and thermodynamic parameters of high complexity associated with biodiesel synthesis. The related transesterification reactions involving plant oils and methanol in a strongly alkaline medium will be simulated using QM/MM multiscale and the empirical valence bond (EVB) method using MOLARIS, Q, and GAUSSIAN. The use-case will use computer cluster and experimental equipment for kinetic studies. The use-case fits the societal challenge Competitive low-carbon energy. The simulations produce a large volume of complex and diverse data including experimental kinetic parameters, trajectories and rheological information that requires new protocols for data storage, sharing, and analysis.

5.5 The Innovativity of the Open Science Design

5.5.1 Service Layers and Data Storage

At present, the various chemistry subdomains carry out their own computational researches like a dispersed archipelago rather than like a networked open system of specialties. The consequences of this situation are as follows:

- Researchers of one subdomain are unable to access facilities and research products of other subdomains (or it is too complicated to use them) with this applying often also to researchers of the same subdomain when this is reasonably large.
- Research facilities are often underdimensioned for the use by the whole community while, because of a too much local nature of the management they are also underutilized.

As a result, the circulation of the produced knowledge (scientific data) is limited and the receptivity for input from other laboratories is insufficient. In the project, thanks to the creation of a collaborative open environment bearing two-way communication between computational and experimental chemistry based on a third generation science gateway, researchers will be able to use data as a common currency for communication within the shared environment. Experimentalists will run their experiments and publish results in the open environment while Computational Chemists will design simulations as a complement to produce new research achievements after analysing this data. In such open environment, computational researchers, on their side, will run simulations whose results can be further checked in experiments. The members of the project will have access to European, national and regional data archives, databases, data centers and data storages using either basic data transfer protocols or advanced B2xx services.[7] The science gateway will have three layers: community, service, and infrastructure access layer.

The community layer will offer social media type services allowing Experimental Chemists to run experiments on remotely available research facilities. This layer will provide to access the submission service to run simulations. It will also support training activities and community building.

The service layer will connect researchers to the research facilities and e-infrastructure resources using microservices managed by a service orchestrator. The set of microservices will contain a data, information, monitoring, resource broker, submission, visualization, etc. service. The prominent innovation will be the data service that will connect Experimental and Computational Chemists through scientific data. Experimental Chemists will use the data service to manage experimental data while Computational Chemists will run simulations through the submission service using the data service. The submission service will support running jobs, pipelines and workflows.

The infrastructure access layer will have two services: computing and data infrastructure access service. The first one will manage access to major computing resources such as cloud, cluster, grid, and supercomputer. The second one will manage data using different types of data resources, such as data archives, databases, data collections, data storages using EUDAT B2xx and MASi services, and major data transfer protocols.

In chemistry, the data life cycle ranges from the upgrade of primary experimental data and fully annotated simulation to fully annotated scientific data requires a data management approach aimed at facilitating reusability and reproducibility using metadata. The large volume of primary data and the diversity of their formats make it difficult to share data among researchers. Moreover, storing and sharing primary (raw) experimental data might not be meaningful because it does not contain information about how it was obtained and processed. Adding metadata to primary data, particularly provenance information, facilitates the sharing of scientific data. Metadata can describe the method and the equipment used, measurement protocol applied, conditions and parameters specified, etc. and can therefore help researchers

[7]The acronym B2xx means "Business to xx" where xx is the beneficiary of the service.

in evaluating the experiment itself and in deciding about further usage of data. The same approach must be followed with simulated data. Similarly to experimental data, significant efforts have been spent to describe computing resources needed, implementation methods used and scientific analysis applied in simulations.

There are a few approaches that support the transparent storage and sharing of scientific simulation data. Markup languages like CML or its derivate MSML offer ontologies for the hierarchical representation of simulation protocols as workflows including relevant input and output data, as well as the analysis. QC-ML and consequently Q5Cost follow a similar tree representation overall focusing more on quantum chemical simulation data. The proposal will build on the experience with MSML and Q5Cost to create a uniform standardized representation of the whole data life cycle ranging from initial experimental data to the analysis of simulation data by feeding the metadata to B2FIND and making it available beyond the closer computational chemistry community. By representation through an XML-based markup language, individual tasks along such community workflows are decoupled from actual implementations, e.g., specific software packages while maintaining the actual purpose of the respective task. For example, the geometry optimization of a given molecule can be accomplished with numerous tools, while the final confirmation should be sufficiently comparable among all implementations. A meta description of such tasks supports the reproducibility and sustainability of scientific protocols in the best possible way.

The discussed researchers' access procedures will allow wider, simplified, and more efficient access to European, national and regional facilities and resources to conduct their research irrespective of the location where they are. This RI will be an open architecture that will serve as transparent basis for future scientific developments inside and outside chemistry. This open architecture will enable connecting further research facilities and resources to extend the outreach outside the project consortium by allowing access to researchers not involved in the project. To further improve research a uniform and standardized data management to handle data ranging from experimental to simulation data will be provided. Moreover, the consistent annotation with provenance and metadata information ensures reusability and reproducibility of scientific results, improving trust into their reliability. This data management solution will allow sharing of information and knowledge between the chemistry and other communities such as Climate and Energy community and between academia and industry.

5.5.2 *Multidisciplinarity, Societal Challenges, Impact and Dissemination*

Important indicators of openness of a project are multidisciplinarity and societal challenges. As to multidisciplinarity, an important theme is the Energy and Climate (with items like energy efficiency and low-carbon energy, which are directly linked to the subject of the book) that has also a key role in targeting also societal needs. In particular joint multidisciplinary research on chemical processes is important in use-case 1 (investigating energetic molecules of potential interest in energy storage/release) and use-case 6 (developing more efficient, cleaner and fuel flexible combustion devices/processes for distributed energy production addressing requirements of the Energy Trilemma (security, equity and sustainability of energy production systems); reducing environmental and health impact of alternative and fossil combustion systems). The same applies to joint multidisciplinary research on Energy efficiency in use-case 2 (studying the photochemistry and reactivity of energetic materials, the laser manipulation of materials, the dynamics, stereo-dynamics and quantum control of elementary chemical processes), use-case 3 (development of technological applications, related to CO_2 abatement (e.g., destruction in electric discharges or by molecular sieves), controlled thermonuclear fusion energy, efficient use of energy in technological applications, such as nuclear fusion by inertial confinement, material science for aerospace and microelectronics applications, plasma-based energy recovery devices), use-case 4 (development of metal complexes for solar devices, efficient energy transfer, determination of electron and energy transfer pathways), use-case 5 (promoting usage of renewable energies by improving storage of renewable energy as carbon neutral fuels), use-case 7 (designing new materials at the nanoscale by combining experimental and numerical results, to improve production of specialized nanomaterials for energy applications), and use-case 8 (improving production of biodiesel fuel and reducing the need for fossil fuels).

As to joint multidisciplinary research on chemical processes for Low-carbon energy it is important in use-case 3 (plasma modeling of applications related to waste treatments (plasma torches, syngas production)), use-case 4 (development of ecological and sustainable catalysts for production of biodegradable plastics from renewable resources to address depletion and exploding costs of fossil resources, climate change and growing landfill sites), use-case 5 (recycle CO_2 by reduction using H_2 to carbon compounds useful for syntheses as well as modeling of the related system based on an accurate prediction of rate coefficients and integration of kinetic equations), and use-case 6 (exploitation of novel energetic molecules that derive from different and locally diverse sources to minimize the CO_2 and pollutant emission).

In Europe, societal challenges are often traced back to collaboration with ESFRI projects. In particular prospective cooperation with ESFRI projects are envisageable in the energy sciences use-cases 5, 6, 7, and 8 (ECCSEL); environmental sciences use-cases 5 and 6 (IAGIOS); and physical sciences use-cases 1, 2, 3, and 4 (IFMIF, ELI, and EuroFEL). The mentioned research cooperation with ESFRI initiatives in facing societal challenges will envisage the development of synergies and complementary capabilities, leading to improved and harmonized services by leveraging on the eight use-cases. This will avoid duplications of facilities and services and will lead to their improved use across Europe. Economies of scale and saving of resources are also realized due to common development and the optimisation of operations. The integration of major research facilities, e-infrastructure resources and of the community knowledge base (collections, archives, structured scientific information, data infrastructures, etc.) will lead to a better management of data collected or produced by these facilities and resources.

Other important characteristics of the openness measures of the project are impact and dissemination that will leverage on the handling the whole data life cycle from primary experimental data to annotated simulation data. Metadata annotation employing a standardized markup format and corresponding ontologies will enable the molecular science community to handle the plenitude of data and formats. A distributed storage infrastructure will host the data and make it available persistently to the community through the usage of EUDAT B2xx and MASi services. Data will be curated by the related use-case as will be specified in an ad hoc agreement. Hence not only the data itself will be available for the scientific community but also the protocols used to generate it, largely improving reproducibility and reusability.

The project will run specific activities to disseminate and exploit projects results. The involvement of SMEs and since long active associations will allow to put on a solid ground such aspect by setting up the strategies to guarantee the maximum impact and sustainability beyond the project lifetime. Moreover, the requirements coming from the project partners will be regularly collected and analyzed by the project technical management to ensure that the requirements are adequately prioritized in the development technical plans, thus maximizing the interoperability with the existing e-infrastructures. Besides, the technical effort to streamline the adoption of the RI products by other communities a sustainable exploitation of the outcomes is only assured if the research communities use them. Any networking strategy addressed to expanding the user base of the RI products among research communities and private companies needs to take into consideration the organizational possibilities and constraints of the research sector. The consortium here counts on the strong support of the different and well-established research institutions plus, as well, of a pool of SMEs which will act as conduit of the new services toward those communities outside the project initiative.

An important level of trans-community services is the one more concerned with the transfer of molecular science achievements to innovation and societal grand challenges. Possible services of this type are as follows:

- Accurate multiscale modeling of smart energy carriers in combustion, energy storage, space missions, and bioinorganic chemistry, using ab initio and empirically parametrized kinetic data;
- Computational design of materials and supramolecular phenomenologies (like clathrate hydrates capture of gases and the properties of ionic liquids), automatic parameterization of molecular dynamics force fields; and
- Handling of extended databases for the investigation of the structure and processes of complex molecular systems relevant to pharmacology, medicinal and biological systems. These services can replace existing services with other ad hoc designed properties.

Finally, another level of vital importance, trans-community services is the one concerned with the management of distributed knowledge to the end of supporting rationalization, dissemination, and education related to CMMST sciences and technologies. At this level self-learning and self-assessing services are considered for both the specific CMMST environment training and the more general educational endeavors of the community for molecular science and technologies-based disciplines able to trigger virtuous learning cycles (in particular those associated with the assemblage, use, and trial-and-error mutual improvement of Learning Objects (LO)s) as fostered by the European Chemistry Thematic Network (ECTN) Association (http://www.expe.ectn-assoc.org/).

Regarding dissemination and exploitation of results the communication activities they will be divided in internal (within the molecular science community) and external (with potential stakeholders). The internal communication will aim at reinforcing cooperation among the community to promote effective synergies. The key to accomplish this goal will be running communication channels among all parties: project partners, user communities, facility and resource providers. The main internal communication channel will be the project website and regular online project, work package, and use-case meetings. The project will also use traditional communication channels such as the ECTN newsletter, published quarterly, the VIRT&L-COMM open access journal to present the RI and use-case achievements. The external communication will focus on disseminating success stories to potential new stakeholders, such as user communities, facility and resource providers, industry partners etc. to raise their awareness about the RI. This will be accomplished through demonstrations, presentations, and publishing in scientific journals. Attending events will also play an important role in outreach activities to new stakeholders. To promote the RI and train its prospective users the project will elaborate a dissemination and training plan to be run in parallel with the usual activities of the project partners (courses, conferences, summer schools, training events, etc.). These activities will be focused also on multidisciplinary research, technology transfer between academia and industry.

Regarding dissemination events, a regular annual workshop will be organized at the Computational Chemistry Conference to raise awareness of the chemistry community about the RI involving facility and technology providers, research and SME partners. The project partners will also present the RI at other chemistry conferences and EGI, EUDAT and PRACE events. They will outline the RI itself, how to use it (focusing on the collaborative activities supported by the RI), and how to extend it. Particular attention will be devoted to the data management highlighting how to use different data formats, how to use metadata and provenance. The project partners will present and demonstrate the use-cases. The project partners will also approach researchers from inside and outside the chemistry community to identify further prospective use-cases to be ported to the RI. Training events. ECTN, Master-UP, and Polymechanon will organize a summer school in every project year on how to use the RI (involving research facilities and e-infrastructure resources providers) for junior researchers and Ph.D. of the Theoretical Chemistry and Computational Modelling (TCCM) ITN JDP. ECTN will define specific learning objectives (LO) for each summer school using the GLOREP archives. ECTNs, Master-UPs, and Polymechanons expertise will guarantee that not only researchers will benefit from training events but also Ph.D. students and SME employees thanks to the use of multimedia technologies.

5.5.3 *User and Service Quality Evaluation*

It is important here to emphasize that the proposed collaborative open science model of the CMMST environment strongly relies on the possibility of evaluating the QoS (quality of service) and the QoU (quality of user) and base on them a credit economy rewarding resource and service providers. This enhances the sustainability of a community by calling both for introducing a metric suited to rank services and for developing tools facilitating collaborative activities by leveraging on the evaluation of QoS and QoU that will be discussed in more detail later on. As already mentioned, in fact, within the collaborative open molecular science model, users not only can get an on-demand allocation of the available compute resources but they can also set common requirements, share in a bottom-up fashion data and programs, complement each other's expertise, boost the activities of their virtual communities, etc. The e-infrastructures developed for that purpose and related software tools are the ground on which various VOs and VRCs originated and performed interdisciplinary collaborations.

At the same time, the adoption of a collaborative model further enhances the competitiveness of the various scientific laboratories thanks to the collaboration developed within the VRE (the already mentioned collaborative competition) by enhancing the complexity of problems affordable (typically the high-level ab initio electronic structure and dynamics calculations, the design of smart energy carriers, innovative materials and biomedical processes, the handling of knowledge for training and education considered in the present project) and the evolution of the quality-based credit system into a business model ensuring sustainability. Preliminary indications on the transdisciplinary performance/research indicators of the impact of the collaborative model have been already pointed out to be accessibility, integrity, and reliability of the collaborative effort (at qualitative level) and number of successful compilations, number of results retrieved and number of feedbacks produced (at quantitative level) as illustrated in Ref. [134].

In this spirit, the open science community is engaged in

- implementing a credit award and redemption system introducing a service-based economy and
- developing a SMEs (academic spinoffs and startups, etc.) driven business model proof of concept.

This activity takes care of providing solutions for

- high-level computing tasks of data processing (e.g., like user-driven workflows and data-driven pipelines),
- simulation workflows and any other stand-alone computing activities and related components, tools, and interfaces built on top of the existing technologies (frameworks and gateways), and
- devising the best solution enabling the concerted use of both the existing users data and applications of the involved communities on the ground of the adoption of the already mentioned QoS and QoU mechanisms.

This is based on a gateway layer that enables

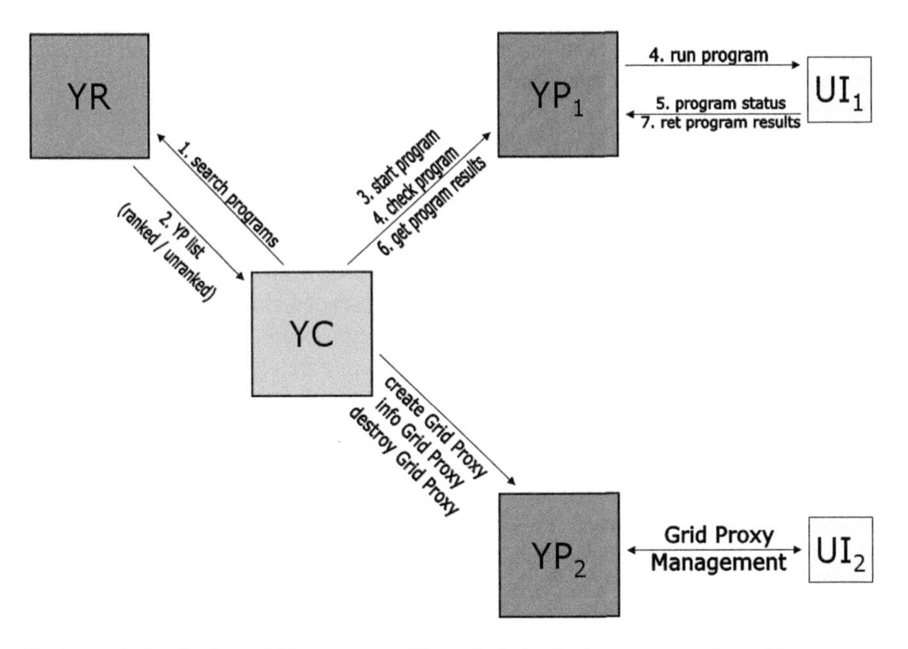

Fig. 5.12 A sketch of the GriF structure and its articulation in Java servers and provider

• users to get proper access to the needed resources of the underlying e-infrastructures and

• software developers to implement a first prototype multiplatform workflow for the underlying use-cases facilitating interoperability among different workflow systems and the user communities by exploiting user and resource ranking in terms of QoS and QoU by incorporating the functionalities of GriF [134] and enabling the ranking and selection of the computing resources to be used.[8]

5.5.4 A Credit Economy

The collaborative model leverages on the promotion of a credit based economy encouraging proactive users to carry out activities (either as work performed within

[8]As illustrated in Fig. 5.12 GriF is a framework made of two Java servers (YC and YR the Consumer and the Registry servers) and one Java client (YP the Provider). The entry points to the computational platforms are the User Interfaces which are able to capture, out of the data supplied by the monitoring sensors of the DCI, the information relevant to properly manage the computational applications of interest and articulate them in sequential, concurrent or alternative quality paths by adopting a service-oriented architecture (SOA) and Web Services. This allows at the same time the guided search of the compute resources on the DCI and the evaluation of the quality of the users (QoU). The computational services provided, are analyzed and used to compose the submission, the monitoring, and the results recollection of molecular science simulations.

the VRE commitments or even as new procurements) useful to the communities to the end of enhancing not only research and innovation but also sustainability. In fact, the objective determination of QoU and QoS quality parameters through extended statistical analyses are utilized for the evaluation of the terms of exchange between the activities carried out and the credits awarded as well as between the credits redeemed and the services or the financial resources provided in return.

This enhances full interoperability of codes, data, tools, and laboratories by encouraging the sharing of packages and results derived from the community endeavors and outcomes thereby exalting the scientific impact of distributed computing. This is also a significant move toward open science, a European project to promote free access to knowledge produced by publicly funded research [151].

A further important element of strategic relevance is the structuring and transforming the credit mechanism into a systematic sustainable market-oriented business model. The goal of this is to deeply involve technology providers and market operators to the end of establishing and maintaining exploitation activities. Key elements of such exploitation activities will include assessment of the optimal business model to adopt in order to transfer related technologies to the market in a way that ensures the widest uptake and profitability as well as business sustainability. Such transfer will leverage both on the gathering of a proper cluster of solid technology providers and on the establishing of a credit mechanism to support the collaborative model adopted by the CMMST environment based on an objective evaluation of the already mentioned terms of exchange in order to ensure that software and services are provided at a professional level with alignment to the established standards.

As mentioned before, the rewarding of the user's contribution to the community activities based on quality evaluation of the services provided is a key added value of the collaborative model. It fosters, in fact, the sharing of expertise and products among the members of the community (especially to the end of tackling multiscale problems of higher complexity for which individual competences are insufficient but including also education for which little scientific recognition is attached and most of the materials are strongly (almost uniquely) linked to the author) in a trustable objective way once the related metrics of monitoring the activities, evaluating their quality and assigning related credits is agreed by the community. It also strongly encourages some members of the community to become proactive service providers by utilizing their own products and those of the other members of the community.

To this end, use is made of GCreS (grid credit system) a tool to reward both the QoU and the QoS it will be possible to assign to the users a congruent amount of credits (according to agreed mechanisms). Such credits, redeemable in terms of a preferential utilization of the community resources (selection of compute systems, DCI services, low- and high-level capabilities, memory size, CPU/wall time, storage capacity plus financial resources) will not only foster collaboration among the members of the community but will also entitle related researchers to participate in the multicompetence teams which apply for the most challenging bids. This will result in an enhancement of the competition among different teams (the so-called competitive collaboration).

Moreover, the use of objective rewarding criteria paves the way to the development of good practices for service provision and market-oriented procedures [135] thanks to the assimilation of service suppliers and proactive users to producers and customers. Service suppliers can offer ordinary and specialistic hardware and software, support the design and development of new algorithms and applications, assist and help the users in running existing packages. On their side, proactive users can produce and validate new datasets, design and develop new grid approaches, disseminate community activities and create, therefore, new possibilities of income for the related scientific area. This ensures that the community not only better accomplishes production work but also feeds new research and development on which grounding future evolution and sustainability for research and innovation.

5.6 Problems

5.6.1 *Qualitative Problems*

1. **Molecular Structure**: Explain how you would use the GAUSSIAN program to calculate the equilibrium positions of the SF_6 molecule. Be sure to specify the method (HF, CCSD, CCSDT, DFT, ...) and the parameters to be used in the input. Now explain how you would calculate the intermolecular potential for $He + SF_6$. Suppose you had a limited budget and could calculate less than 1000 points for intermolecular potential. Explain how you would pick those point. What analytical forms and procedures would you use to accurately fit the intermolecular potential?

5.6.2 *Quantitative Problems*

1. $He + SF_6$ **potential**: Assume the F atoms are rigidly fixed to the central S atom at their equilibrium positions. Construct a reasonable intermolecular potential for $He + SF_6$ by using the diatomics in molecules procedure. Use simple Lennard-Jones potentials for each of the $He + S$ and $He + F$ interactions $V^{HeS}(r_{HeS})$ and $V^{HeF}(r_{HeF})$, respectively. Explain how one could use molecular structure calculations to more accurately fix the 4 Lennard-Jones parameters $(\epsilon_{HeS}, \sigma_{HeS}, \epsilon_{HeF}, \sigma_{HeF})$.
2. **Inelastic Scattering**: The infinite Order Sudden Approximation is quite useful for calculating approximate inelastic nonreactive results. This approximation is valid when the total energy is large compared to the rotational energy spacing. It is equivalent to holding the orientation angles $[\gamma, \phi]$ fixed between the atom and the polyatomic molecule during the collision. That is the molecule does not significantly rotate during the collision process. Use the JWKB phase shift program to calculate the phase shifts $\eta_l(\gamma, \phi)$ for the $He + SF_6$ molecule assuming it is

a rigid rotor. The orbital angular momentum quantum number should vary from 0 to 100 for each angle. Then calculate the integrated cross section by averaging over the orientation angles

$$\sigma = \frac{1}{2} \int_{-1}^{1} \int_{0}^{2\pi} \sigma(\gamma, \phi)\, d\cos(\gamma)\, d\phi \tag{5.26}$$

where

$$\sigma(\gamma, \phi) = \frac{4\pi}{k^2} \sum_{l=0}^{l=100} (2l+1) \sin^2 \eta_l(\gamma, \phi) \tag{5.27}$$

By the way state-to-state differential cross sections, generalized cross sections, as well as bulk properties such as spectral line broadening, diffusion, viscosity, and virial coefficient can be calculated as well. This problem is very similar to Problem 4.4 in Chap. 4.

Appendix

A.1 Vectors and Matrices Spaces and Operators

$|\psi\rangle$ and $\langle\psi|$ (with the former being a row matrix of ψ values and the latter a column matrix of its complex conjugated ψ^*) denote the "bra" and "ket" vectors of the braket space of popular use in quantum chemistry. Symbols denoting matrices are written in bold and the determinant of matrix **A** is written as $|\mathbf{A}|$ or det (**A**). The transposed matrix **A** is written as $\mathbf{A}^{\mathrm{T}}$, the adjugate matrix **A** is written as $\mathbf{A}^{+}$, and the inverse matrix **A** is written as $\mathbf{A}^{-1}$. The unit matrix (**1**) is the matrix made of all zeroes but diagonal ones which have value 1, while the matrix made of all zeroes is the zero matrix **0**.

A vector space of dimension n is called V^n with V^2 corresponding to a plane (say x,y) and V^3 corresponding to a three-dimensional space (say x, y, z). Basis sets of the vector space V^n are usually denoted as $\{e_i\}^n$. In the case of V^3, one has usually $\{e_i\}^3 = (|i\rangle, |j\rangle, |k\rangle)$. The Function space F^n is the analogue of the vector space consisting of n linearly independent basis functions ϕ_i ($\{\phi_i\}^n$) owing to the fact that one replaces the variable x with its function ϕ_i (strictly speaking this is a continuous transformation but it can be considered as the limit $\Delta x \rightarrow 0$ of the function representation in steps of Δx corresponding to V^∞). This vector–function equivalence allows easier manipulations of functions especially for compute purposes.

Vectors and functions are often constructed from sets of basis vectors or functions like the above-mentioned $\{e_i\}^n$ and $\{\phi_i\}^n$. Operators acting on vectors and matrices are marked by the "hat." Special operators are the direct sum ($\oplus$) and product ($\otimes$). Of particular interest for quantum chemistry are the Nabla (∇) defined as

$$\frac{\partial}{\partial x}|i\rangle + \frac{\partial}{\partial y}|j\rangle + \frac{\partial}{\partial z}|k\rangle, \tag{A.1}$$

and the Laplacian (Δ or ∇^2) defined as

$$\frac{\partial^2}{\partial x^2} + \frac{\partial^2}{\partial y^2} + \frac{\partial^2}{\partial x^2}. \tag{A.2}$$

A. Laganà and G. A. Parker (eds.), *Chemical Reactions*, Theoretical Chemistry and Computational Modelling, https://doi.org/10.1007/978-3-319-62356-6

Given the vectors $|a\rangle$ and $|b\rangle$, their scalar product is given by $\sum_i a_i^* \cdot b_i$ (in matrix notation $\mathbf{A}^+ \cdot \mathbf{B}$ in which the matrices $\mathbf{A}$ and $\mathbf{B}$ must correspond to the same basis $\{e_i\}$ i.e., $|a\rangle = c_{a_1} \cdot |e_1\rangle + c_{a_2} \cdot |e_2 + \ldots c_{a_n} \cdot |e_n\rangle = \mathbf{E} \cdot \mathbf{C}_a$ and $|b\rangle = c_{b_1} \cdot |e_1\rangle + c_{b_2} \cdot |e_2 + \ldots c_{b_n} \cdot |e_n\rangle = \mathbf{E} \cdot \mathbf{C}_b$ with $\mathbf{E}$ representing the basis, $\mathbf{C}_a$ and $\mathbf{C}_b$ the column vectors of coefficients) corresponding for a function to the integral (a sum with an infinitesimal step) i.e., $\langle \phi_1 | \phi_2 \rangle$ corresponds to $\int \phi_1^* \cdot \phi_2 d\tau$. More in general, the scalar product of two matrices, say $\mathbf{A}$ and $\mathbf{B}$, is a third matrix $\mathbf{C}$ whose elements are defined as $c_{i,j} = \sum_k a_{i,k} b_{k,j}$. Scalar products are involved in important matrix operations like those needed for carrying out a linear transformation from one basis to another. This is performed by multiplying the matrix expressed in the previous basis, say $\mathbf{A}$, by a transformation matrix $\mathbf{T}$ to generate the matrix $\mathbf{B}$ expressed in the new basis $\mathbf{B} = \mathbf{A} \cdot \mathbf{T}$ (or equivalently $\mathbf{BT}^{-1} = \mathbf{A}$).

A.2 Derivative Proof

Let $f(x)$ be real and continuous as well as its derivatives, for all x an element of the reals, then

$$\frac{1}{f^2(x)} \frac{d}{dx} \left(f^2(x) \frac{d}{dx} \Psi(x) \right) = \frac{1}{f^2(x)} \left[\frac{d}{dx} \left(f^2(x) \Psi'(x) \right) \right] \tag{A.3}$$

$$= \Psi''(x) + 2 \frac{f'(x)}{f(x)} \Psi'(x) \tag{A.4}$$

$$= \left[\frac{d^2}{dx^2} + \frac{f'(x)}{f(x)} \frac{d}{dx} \right] \Psi(x) \frac{f''(x)}{f(x)} \tag{A.5}$$

and

$$\frac{1}{f(x)} \frac{d^2}{dx^2} [f(x)\Psi(x)] = \frac{1}{f(x)} \frac{d}{dx} \left[f(x)\Psi'(x) + f'(x)\Psi(x) \right] \tag{A.6}$$

$$= \Psi''(x) + 2 \frac{f'(x)}{f(x)} \Psi'(x) + \frac{f''(x)}{f(x)} \Psi(x) \tag{A.7}$$

$$= \left[\frac{d^2}{dx^2} + 2 \frac{f'(x)}{f(x)} \frac{d}{dx} + \frac{f''(x)}{f(x)} \right] \Psi(x) \tag{A.8}$$

comparing Eqs. A.5 and A.8

$$\frac{1}{f^2(x)} \frac{d}{dx} \left(f^2(x) \frac{d}{dx} \right) = \frac{d^2}{dx^2} + 2 \frac{f'(x)}{f(x)} \frac{d}{dx} = \frac{1}{f(x)} \frac{d^2}{dx^2} f(x) - \frac{f''(x)}{f(x)} \tag{A.9}$$

Now letting $x = r$ and $f(r) = r$, we have

$$\frac{1}{r^2}\frac{d}{dr}\left(r^2\frac{d}{dr}\right) = \frac{d^2}{dr^2} + \frac{2}{r}\frac{d}{dr} = \frac{1}{r}\frac{d^2}{dr^2}r \tag{A.10}$$

which is the verification of Eq. 2.22. Likewise, if we let $x = \rho$ and $f(\rho) = \rho^{5/2}$, we have

$$-\frac{\hbar^2}{2\mu}\frac{1}{\rho^5}\frac{d}{d\rho}\left(\rho^5\frac{d}{d\rho}\right) = -\frac{\hbar^2}{2\mu}\frac{1}{\rho^{5/2}}\frac{d^2}{d\rho^2}\rho^{5/2} + \frac{15}{8\mu\rho^2} \tag{A.11}$$

which is the verification of the formulation of T_ρ in Eq. 4.2.3.

A.3 Partial Wave Expansion of the Elastic Scattering Wavefunction

Determination of the coefficients $\tilde{A}_l$ in the expansion in partial waves, Eq. (2.59):

$$\Psi_{inc}(\mathbf{r}) = e^{ikz} = e^{ikrt} = \frac{1}{r}\sum_{l=0}^{\infty}\tilde{A}_l\tilde{\xi}_l(r)P_l(t) \quad \text{with } t = \cos\theta \tag{A.12}$$

Multiplying both sides of the equation by $P_l(t)$, integrating with respect to the cosine of the angle (t), and using the orthogonality of the Legendre polynomials:

$$\int_{-1}^{1} P_l(t)P_{l'}(t)\mathrm{d}t = \frac{2}{2l+1}\delta_{ll'} \tag{A.13}$$

we obtain

$$\int_{-1}^{1} e^{ikrt}P_l(t)\mathrm{d}t = \frac{1}{r}\tilde{A}_l\tilde{\xi}_l(r)\frac{2}{2l+1}. \tag{A.14}$$

Integrating the left-hand side by parts twice, this equation becomes

$$\int_{-1}^{1} e^{ikrt}P_l(t)\mathrm{d}t = \frac{1}{ikr}\left[e^{ikrt P_l}(t)\right]_{-1}^{1}$$
$$-\frac{1}{ikr}\left\{\frac{1}{ikr}\left[e^{ikrt}P_l'(t)\right]_{-1}^{1} - \frac{1}{ikr}\int_{-1}^{1} e^{ikrt}P_l''(t)\mathrm{d}t\right\}$$

asymptotically $(kr \to \infty)$, we can neglect the terms after the first in the right-hand side and recalling that $P_l(\pm 1) = (\pm 1)^l$, we have

$$\int_{-1}^{1} e^{ikrt P_l}(t) \overset{r\to\infty}{\sim} \frac{1}{ikr}\left(e^{ikr} - (-1)^l e^{-ikr}\right) = i^l\frac{2}{kr}\sin(kr - l\pi/2) \tag{A.15}$$

where we have used the relation

$$\sin z = \frac{e^{iz} - e^{-iz}}{2i}. \tag{A.16}$$

Substituting then (A.15) into (A.14) and comparing with the asymptotic form (2.63), we obtain the result

$$\tilde{A}_l = (2l + 1)i^l/k. \tag{A.17}$$

A.4 Elimination method

```
Irig1 REPEAT FOR GOING FROM 1 TO n-1
  VNORM=vara (irig1,irig1)
  varb(irig1)=varb(irig1)/VNORM
  REPEAT FOR ICOL GOING TO BE A n irig1
    vara (irig1, ICOL) = vara (irig1, ICOL) / VNORM
  END REPEAT ICOL
  REPEAT FOR irig2 GOING TO BE A n +1 irig1
    varb (irig2) = varb (irig2) / VNORM-varb (irig1)
    REPEAT FOR ICOL GOING TO BE A n irig2
       vara(irig2,ICOL)=vara(irig2,ICOL)/VNORM-vara(irig1,ICOL)
    END REPEAT ICOL
  END REPEAT irig2
END REPEAT irig1

REPEAT FOR irig1 GOING TO PASS n 1 -1
   sum = varb (irig1)
   IF (irig1.ne.n) THEN
       REPEAT FOR ICOL GOING TO BE A n +1 irig1 OF STEP -1
         sum = sum - vara (irig1, ICOL)
       END REPEAT ICOL
   END IF
   varX (irig1) = sum
END REPEAT irig1
```

One sees immediately that in this simple form, the algorithm can give us problems because there are situations where the VNORM the variable is zero or nearly zero. The above procedure works (stable) well if the matrix is diagonally dominate or positive definite.

A better procedure is to use partial pivoting which selects the largest absolute value of the column which generally reduces roundoff error. The modified pseudo code is as follows:

```
Irig1 REPEAT FOR GOING FROM 1 TO n
icol1 (irig1) = irig1
END REPEAT irig1
Irig1 REPEAT FOR GOING FROM 1 TO n-1
  amax = vara (irig1, irig1)
  imax = irig1
  REPEAT FOR ICOL GOING TO BE A n +1 irig1
    IF (ABS (vara (irig1, ICOL)). Gt.ABS (amax)) THEN
      amax = vara (irig1, ICOL)
      imax = ICOL
    END IF
  END REPEAT ICOL
  VNORM = amax
  icol1 (imax) = icol1 (irig1)
  icol1 (irig1) = imax
  vara (irig1, imax) = vara (irig1, icol1 (imax))
  vara (irig1, icol1 (imax)) = amax
  VNORM = vara (irig1, irig1)
  varb (irig1) = varb (irig1) / VNORM
  REPEAT FOR ICOL GOING TO BE A n irig1
    vara (irig1, ICOL) = vara (irig1, ICOL) / VNORM
  END REPEAT ICOL
  REPEAT FOR irig2 GOING TO BE A n +1 irig1
    varb (irig2) = varb (irig2) / VNORM-varb (irig1)
    REPEAT FOR ICOL GOING TO BE A n irig2
       vara(irig2,icol1(ICOL))=vara(irig2,icol1(ICOL))/VNORM
               -vara(irig1,icol1(ICOL))
    END REPEAT ICOL
  END REPEAT irig2
END REPEAT irig1

REPEAT FOR irig1 GOING TO PASS n 1 -1
   sum = varb (irig1)
   SE (irig1.ne.n) THEN
       REPEAT FOR ICOL GOING TO BE A n +1 irig1 OF STEP -1
         sum = sum - vara (irig1, icol1 (ICOL))
       END REPEAT ICOL
   END IF
   varX (irig1) = sum
END REPEAT irig1
```

References

1. D.M. Hirst, *A Computational Approach to Chemistry* (Blackwell Scientific Publications, Oxford, 1990)
2. Numerical recipes (http://www.nr.com/oldverswitcher.html)
3. M.S. Child, *Semiclassical Mechanics with Molecular Applications* (Oxford Press Inc., New York, 2005)
4. E. Garcia, A. Laganà, Diatomic potential functions for triatomic scattering. Mol. Phys. **56**, 621–627 (1985)
5. C.J. Joachain, *Quantum Collision Theory* (North-Holland Physics Publishing, Amsterdam, 1975)
6. S. Brandt, H.D. Dahmen, *The Picture Book of Quantum Mechanics* (Wiley, New York, 1985)
7. L. Landau, E. Lifshitz, *Physical-Theoretical Mechanics* (Editori Riuniti, Rome, 1976)
8. N.F. Mott, H.S.W. Massey, *The Theory of Atomic Collisions* (Oxford University Press, Oxford, 1965)
9. M. Abramowitz, I.A. Stegun, *Handbook of Mathematical Functions: With Formulas, Graphs, and Mathematical Tables* (Dover Publications, New York, 2010)
10. A. Schawlow, T.W. Hnsch, G.W. Series, The spectrum of atomic hydrogen. Sci. Am. **240**, 94 (1979)
11. I.N. Levine, *Quantum Chemistry* (Prentice-Hall Inc., New Jersey, 1991)
12. R.G. Newton, *Theory of Scattering Waves and Particles* (McGraw-Hill Book Company, New York, 1966)
13. S. Flügge, *Practical Quantum Mechanics* (Springer, New York, 1970)
14. R. McWeeny, *Methods of Molecular Quantum Mechanics*, 2nd edn. (Academic Press, London, 1992)
15. F. Jensen, *Introduction to Computational Chemistry*, 2nd edn. (Wiley, New York, 2007)
16. B.M. Rode, T.S. Hofer, M.D. Kugler, *The basic of Theoretical and Computational Chemistry* (Wiley-Vch GMBH, New York, 2007). ISBN 978-3-527-31773-8
17. Quantum Chemistry packages list: http://en.wikipedia.org/wiki/List_of_quantum_chemistry_and_solid_state_physics_software
18. W.K. Hastings, Monte Carlo sampling methods using Markov chains and their application. Biometrika **57**(1), 97 (1970)
19. M. Caffarel, P. Claverie, J. Chem. Phys. **88**(2), 1088 (1988)
20. A. Szabo, N.S. Ostlund, *Modern Quantum Chemistry: Introduction to Advanced Electronic Structure Theory* (Dover Publications, New York, 1996)
21. C.C.J. Roothaan, New developments in molecular orbital theory. Rev. Mod. Phys. **23**, 69–89 (1951)

A. Laganà and G. A. Parker (eds.), *Chemical Reactions*, Theoretical Chemistry and Computational Modelling, https://doi.org/10.1007/978-3-319-62356-6

22. B. Roos, Chem. Phys. Lett. **15**, 153 (1972)
23. W.J. Hehre, L. Radom, P. von R. Schleyer, J.A. Pople, *Ab Initio Molecular Orbital Theory* (Wiley, New York, 1986)
24. P.J. Kuntz, E.M. Nemeth, J.C. Polanyi, S.D. Rosner, C.E. Young, J. Chem. Phys. **44**, 1168 (1966)
25. G.C. Schatz, Fitting potential energy surfaces, in *Reaction and Molecular Dynamics*, Lecture Notes in Chemistry, ed. by A. Laganà, A. Riganelli (Springer, Berlin, 2000), p. 15
26. J.N. Murrell, S. Carter, S.C. Farantos, P. Huxley, A.J.C. Varandas, *Molecular Potential Energy Functions* (Wiley, Chichester, 1984)
27. E. Garcia, A. Laganà, A new bond order functional form for triatomic molecules: a fit of the BeFH potential energy. Mol. Phys. **56**, 629–639 (1985)
28. A. Aguado, M. Paniagua, J. Chem. Phys. **96**, 1265 (1992)
29. B.J. Braams, J.M. Bowman, Permutationally invariant potential energy surfaces in high dimensionality. Int. Rev. Phys. Chem. **28**, 4 (2009)
30. K.C. Thompson, M.A. Collins, J. Chem. Soc. Faraday Trans. **93**, 871 (1997)
31. P. Lancaster, K. Salkauskas, *Curve and Surface Fitting: An Introduction* (Academic, London, 1986)
32. P.J. Kuntz, in *Atom Molecule Collision Theory*, ed. by R.B. Bernstein (Plenum, New York, 1979), p. 79
33. A.J.C. Varandas, A.I. Voronin, Mol. Phys. **85**, 497 (1995)
34. E. Garcia, C. Sanchez, A. Rodriguez, A. Lagana, A MEP-MPE potential energy surface for the Cl + CH4 reaction. Int. J. Quantum Chem. **106**, 623–630 (2006)
35. A. Laganà, A rotating bond order formulation of the atom diatom potential energy surface. J. Chem. Phys. **95**, 2216–2217 (1991)
36. M. Verdicchio, Atmospheric reentry calculations and extension of the formats of quantum chemistry data to quantum dynamics, Master Thesis, University of Perugia, 2009
37. D. Wang, J.R. Stallcop, W.M. Huo, C.E. Dateo, D.W. Schwenke, H. Partridge, J. Chem. Phys. **118**, 2186–2189 (2003)
38. B.R.L. Galvão, A.J.C. Varandas, J. Phys. Chem. A **113**, 14424–14430 (2009)
39. A. Laganà, E. Garcia, J. Chem. Phys. **103**, 5410 (1995)
40. S. Rampino, J. Phys. Chem. A **120**(27), 4683–4692 (2016)
41. T.W. Hansch, A.L. Schawlow, G.W. Series, The spectrum of atomic hydrogen. Sci. Am. **240**, 94 (1979)
42. T.J. Zielinski, E. Harvey, R. Sweeney, D.M. Hanson, Quantum states of atoms and molecules. J. Chem. Educ. **82**(12), 1880 (2005)
43. R.J. Bartlett, M. Musial, Coupled-cluster theory in quantum chemistry. Rev. Mod. Phys. **79**, 91 (2007)
44. H.D. Meyer, F. Gatti, G.A. Worth, in *Multidimensional Quantum Dynamics: MCTDH Theory and Applications*, ed. by H.D. Meyer, F. Gatti, G.A. Worth (Wiley, Chichester, 2009)
45. D. Skouteris, A. Laganà, in, Lecture Notes in Computer Science **6784**, 442–452 (2011)
46. D. Skouteris, J.F. Castillo, D.E. Manolopulos, ABC: a quantum reactive scattering program. Comp. Phys. Commun. **133**, 128135 (2000)
47. D. Skouteris, D. Pacifici, L. Laganà, Time dependent wavepacket calculations for the $N(^4S)$ + $N_2(^1\Sigma_g^+)$ system on a LEPS surface: inelastic and reactive probabilities. Mol. Phys. **102**(21–22), 2237–2248 (2004)
48. M. Hankel, M. Smith, C. Sean, S.K. Gray, G.G. Balint-Kurti, DIFFREALWAVE: a parallel real wavepacket code for the quantum mechanical calculation of reactive state-to-state differential cross sections in atom plus diatom collisions. Comput. Phys. Commun. **179**(8), 569–578 (2008)
49. R.T. Pack, G.A. Parker, J. Chem. Phys. **87**, 3888 (1987)
50. U. Manthe, in *Direct calculations of reaction rates in Reaction and Molecular Dynamics*, ed. by A. Laganà, A. Riganelli (Springer, Berlin, 2000), p. 130
51. M. Beck, A. Jakle, G. Worth, H.D. Meyer, The multiconfiguration time dependent Hartree (MCTDH) method: a highly efficient algorithm for propagating wavepackets. Phys. Rep. **324**, 15 (2000)

52. G.G. Balint-Kurti, Time dependent quantum approaches to chemical reactivity, in *Reaction and Molecular Dynmics*, ed. by A. Laganà, A. Riganelli (Springer, Berlin, 2000), p. 74; G.G. Balint-Kurti, A. Palov, *Theory of Molecular collisions*, (Royal Society of Chemistry, 2015)
53. J.C. Light, T. Carrington, Adv. Chem. Phys. **114**, 263 (2000)
54. J. Crawford, G.A. Parker, State-to-state three-atom time-dependent reactive scattering in hyperspherical coordinates. J. Chem. Phys. **138**, 054313 (2013). A version of the code is available for distribution
55. D. Chase, Phys. Rev. **104**, 838 (1956)
56. R.T. Pack, J. Chem. Phys. **60**, 653 (1974)
57. T. Kato, On the adiabatic theorem of quantum mechanics. J. Phys. Soc. Jpn. **5**(6), 435439 (1950)
58. J.M. Bowman, Approximate time independent methods for polyatomic reactions, in Lecture Notes in Chemistry, ed. by A. Laganà, A. Riganelli (Springer, Berlin, 2000), p. 101
59. J.R. Schmidt, P.V. Parandekar, J.C. Tully, Mixed quantum-classical equilibrium: surface hopping. J. Chem. Phys. **129**, 044104 (2008)
60. A. Abedi, N.T. Maitra, E.K.U. Gross, Phys. Rev. Lett. **105**, 123002 (2010)
61. W.H. Miller, The semiclassical initial value representation: a potentially practical way for adding quantum effects to classical molecular dynamics simulations. J. Phys. Chem. A **105**, 2942–2955 (2001)
62. H. Jeffreys, Proc. Lond. Math. Soc. **23**, 428 (1925)
63. G. Wentzel, Z. Phys. **38**, 518 (1926)
64. H.A. Kramers, Z. Phys. **39**, 828 (1926)
65. L. Brillouin, J. Phys. **7**, 353 (1926)
66. R.A. Marcus, Chem. Phys. Lett. **7**, 525 (1970)
67. L. Ciccarelli, E. Garcia, A. Laganà, A quantum mechanical test of the DIM surface of the Li + HCl semi-empirical surface. Chem. Phys. Lett. **120**, 75–79 (1985)
68. D.R. Herschbach, Y.T. Lee, J.C. Polanyi, Nobel lecture december 8, 1986, from Nobel lectures, in *Chemistry 1981-1990*, ed. by T. Frängsmyr, B.G. Malmström (World Scientific Publishing Co., Singapore, 1992)
69. G.C. Schatz, A. Kuppermann, Quantum mechanical scattering for three dimensional atom plus diatom systems: II accurate cross sections for H + H_2. J. Chem. Phys. **66**, 4668 (1977)
70. A. Laganà, N. Faginas Lago, R. Gargano, P.R.P. Barreto, On the semiclassical initial value calculation of thermal rate coefficients for the N + N2 reaction. J. Chem. Phys. **125**, 114311–114317 (2006)
71. A. Bar-Nun, A. Lifshitz, J. Chem. Phys. **47**, 2878 (1967)
72. R.A. Back, J.Y.P. Mui, J. Phys. Chem. **66**, 1362 (1962)
73. R. Lyon, Can. J. Chem. **50**, 1437 (1972)
74. D. Skouteris, O. Gervasi, A. Laganà, Non-Born-Oppenheimer MCTDH calculations on the confined H_2^+ molecular ion. Chem. Phys. Lett. **500**(1–3), 144–148 (2010)
75. http://www.molcas.org/
76. http://www.molpro.net/
77. M.W. Schmidt, K.K. Baldridge, J.A. Boatz, S.T. Elbert, M.S. Gordon, J.H. Jensen, S. Koseki, N. Matsunaga, K.A. Nguyen, S. Su, T.L. Windus, M. Dupuis, J.A. Montgomery, J. Comput. Chem. **14**, 1347 (1993), GAMESS-US see http://www.msg.ameslab.gov/gamess/
78. http://www.gaussian.com/
79. M. Valiev, E.J. Bylaska, N. Govind, K. Kowalski, T.P. Straatsma, H.J.J. van Dam, D. Wang, J. Nieplocha, E. Apra, T.L. Windus, W.A. de Jong, NWChem: a comprehensive and scalable open-source solution for large scale molecular simulations. Comput. Phys. Commun. **181**, 1477 (2010), http://www.nwchem-sw.org/index.php/Main_Page
80. R. Ahlrichs, M. Bar, M. Haser, H. Horn, C. Kolmel, Electronic structure calculations on workstation computers: the program system Turbomole. Chem. Phys. Lett. **162**(3), 165–169 (1989)
81. F. Neese, The ORCA program system. WIREs Comput. Mol. Sci. **2**, 7378 (2012)
82. http://www.hondoplus.com/

83. R.D. Amos, J.E. Rice, *CADPAC, The Cambridge Analytical Derivatives Package* (Cambridge, England, 1987), http://www-theor.ch.cam.ac.uk/software/cadpac.html
84. D. Cappelletti, F. Pirani, B. Bussery-Honvault, L. Gomez, M. Bartolomei, A bond-bond description of the intermolecular interaction energy: the case of weakly bound N_2H_2 and N_2N_2 complexes. Phys. Chem. Chem. Phys. **10**(29), 4281–4293 (2008)
85. F. Pirani, D. Cappelletti, G. Liuti, Range, strength and anisotropy of intermolecular forces in atom-molecule systems: an atom-bond pairwise additivity approach. Chem. Phys. Lett. **350**(3), 286–296 (2001)
86. A. Laganà, A bond order approach to a process oriented fitting of potential energy surfaces. VIRT&L-COMM.10.2016.2
87. F. Pirani, S. Brizi, L.F. Roncaratti, P. Casavecchia, D. Cappelletti, F. Vecchiocattivi, Beyond the Lennard-Jones model: a simple and accurate potential function probed by high resolution scattering data useful for molecular dynamics simulations. Phys. Chem. Chem. Phys. **10**(36), 5489–503 (2008)
88. A. Laganà, G. Ochoa de Aspuru, E. Garcia, J. Phys. Chem. **99**, 17139 (1995)
89. G. Ochoa de Aspuru, D.C. Clary, New potential energy function for four-atom reactions. Application to OH + H_2. J. Phys. Chem. **102**(47), 96319637 (1998)
90. A. van der Avoird, P.E.S. Wormer, A.P.J. Jansen, J. Chem. Phys. **84**, 1629–1635 (1986)
91. J.R. Stallcop, H. Partridge, Chem. Phys. Lett. **281**, 212–220 (1997)
92. D. Capelletti, F. Vecchiocattivi, F. Pirani, E.L. Heck, A.S. Dickinson, Mol. Phys. **93**, 485–499 (1988)
93. V. Aquilanti, M. Bartolomei, D. Cappelletti, E. Carmona-Novillo, F. Pirani, J. Chem. Phys. **117**, 615–627 (2002)
94. L. Gomez, B. Bussery-Honvault, T. Cauchy, M. Bartolomei, D. Cappelletti, F. Pirani, Chem. Phys. Lett. **445**, 99–107 (2007)
95. R. Hellmann, Mol. Phys. **111**, 387–401 (2013)
96. D. Cappelletti, F. Pirani, B. Bussery-Honvault, L. Gomez, M. Bartolomei, Phys. Chem. Chem. Phys. **10**, 4281–4293 (2008)
97. L. Pacifici, M. Verdicchio, N. Faginas Lago, A. Lombardi, A. Costantini, J. Comput. Chem. **34**, 2668–2676 (2013)
98. Y. Paukku, K.R. Yang, Z. Varga, D.G. Truhlar, J. Chem. Phys. **139**, 044309 (2013)
99. Y. Paukku, K.R. Yang, Z. Varga, D.G. Truhlar, J. Chem. Phys. **140**, 019903 (2014)
100. J.D. Bender, S. Doraiswamy, D.G. Truhlar, G.V. Candler, J. Chem. Phys. **140**, 054302 (2014)
101. N. Parsons, D.A. Levin, A.C.T. van Duin, T. Zhu, J. Chem. Phys. **141**, 234307 (2014)
102. J.D. Bender, P. Valentini, I. Nompelis, Y. Paukku, Z. Varga, D.G. Truhlar, T. Schwartzentruber, G.V. Candler, J. Chem. Phys. **143**, 054304 (2015)
103. http://www.giss.nasa.gov/tools/molscat/
104. G.D. Billing, Chem. Phys. Lett. **76**, 178 (1980)
105. M. Cacciatore, A. Kurnosov, A. Napartovich, J. Chem. Phys. **123**, 174315 (2005)
106. E. Garcia, A. Laganà, F. Pirani, M. Bartolomei, M. Cacciatore, A. Kurnosov, On the efficiency of collisional $O_2 + N_2$ vibrational energy exchange. J. Phys. Chem. B **120**, 1476–1485 (2016)
107. S. Rampino, N. Faginas Lago, A. Laganà, F. Huarte-Larrañaga, An extension of the grid empowered molecular simulator GEMS to quantum reactive scattering. J. Comput. Chem. **33**, 708714 (2012)
108. A. Laganà, E. Garcia, A. Paladini, P. Casavecchia, N. Balucani, The last mile of molecular reaction dynamics virtual experiments: the case of the OH($N=1-10$)+CO($j=0-3$) reaction. Faraday Discuss. Chem. Soc. **157**, 415–436 (2012)
109. L. Verlet, Computer experiments on classical fluids. Phys. Rev. **159**, 98 (1967)
110. W.L. Hase, R.J. Duchovic, X. Hu, A. Komornicki, K.F. Lim, D. Lu, G.H. Peslherbe, K.N. Swamy, S.R. van de Linde, A.J.C. Varandas, H. Wang, R.J. Wolf, Chemical dynamics software and simulation system (CDSSIM system). QCPE Bull. **16**, 43 (1996), https://cdssim.chem.ttu.edu/nav/htmlpages/licensemenu.jsp
111. M. Ceotto, S. Atahan, S. Shin, First principles semiclassical initial value representation molecular dynamics. Phys. Chem. Chem. Phys. **11**, 3861–3867 (2009)

112. W. Smith, T.R. Forester, DL_POLY2 a general purpose parallel molecular dynamics simulation package. J. Mol. Graph. **14**(3), 136–141 (1996)
113. S. Pronk, S. Pll, R. Schulz, P. Larsson, P. Bjelkmar, R. Apostolov, M. Shirts, J. Smith, P. Kasson, D. van der Spoel, B. Hess, E. Lindahl, GROMACS 4.5: a high-throughput and highly parallel open source molecular simulation toolkit. Bioinformatics **29**(7), 845–854 (2013)
114. J.C. Phillips, R. Braun, W. Wang, J. Gumbart, E. Tajkhorshid, E. Villa, C. Chipot, R.D. Skeel, L. Kale, K. Schulten, Scalable molecular dynamics with NAMD. J. Comput. Chem. **26**, 1781 (2005)
115. O. Trott, A.J. Olson, AutoDock Vina: improving the speed and accuracy of docking with a new scoring function, efficient optimization and multithreading. J. Comput. Chem. **31**, 455 (2010)
116. CADDSuite, http://www.ballview.org/caddsuite
117. Flexx, http://www.biosolveit.de/FlexX/
118. K.J. Bowers, E. Chow, H. Xu, R.O. Dror, M.P. Eastwood, B.A. Gregersen, J.L. Klepeis, I. Kolossvry, M.A. Moraes, F.D. Sacerdoti, J.K. Salmon, Y. Shan, D.E. Shaw, Scalable algorithms for molecular dynamics simulations on commodity clusters, in *Proceedings of the ACM/IEEE Conference on Supercomputing (SC06)* (IEEE, New York, 2006)
119. W.D. Cornell, P. Cieplak, C.I. Bayly, I.R. Gould, K.M. Merz Jr., D.M. Ferguson, D.C. Spellmeyer, T. Fox, J.W. Caldwell, P.A. Kollman, A second generation force field for the simulation of proteins, nucleic acids, and organic molecules. J. Am. Chem. Soc. **117**, 5179 (1995)
120. S. Dietrich, I. Boyd, Scalar and parallel optimized implementation of the direct simulation Monte Carlo method. J. Comput. Phys. **126**, 328 (1996)
121. G.A. Bird, *Molecular Gas Dynamics* (Clarendon, Oxford, 1976)
122. G.A. Bird, *Molecular Gas Dynamics and the Direct Simulation of Gas Flows* (Claredon, Oxford, 1994)
123. H.S. Tsein, Superaerodynamics, mechanics of rarefied gases. J. Aerosp. Sci. **13**, 342 (1946)
124. C.-D. Munz, M. Auweter-Kurtz, S. Fasoulas et al., Coupled particle-in-cell and direct simulation Monte Carlo method for simulating reactive plasma flows. Comptes Rendus Mecanique **342**(10–11), 662 (2014)
125. The European Grid Initiative (EGI), http://www.egi.eu/infrastructure. Accessed 19 Feb 2015
126. https://wiki.egi.eu/wiki/VT_Towards_a_CMMST_VRC
127. A. Voss, R. Procter, Virtual research environments in scholarly work and communications. Libr. Hi Tech **27**(2), 174 (2009)
128. A. Laganà, A. Riganelli, O. Gervasi, On the structuring of the Computational Chemistry Virtual Organization COMCPHEM, Lecture Notes in Computer Science (2006), pp. 665–674
129. A. Laganà, DRAG: a cluster of spinoffs for grid and cloud services. VIRT&L-COMM.5.2014.16
130. A. Capriccioli, Report on PROGEO progress. VIRT&L-COMM.8.2015.2
131. A. Laganà, The Molecular science community for open science. VIRT&L-COMM.9.2016.6
132. A. Laganà, C. Manuali, L. Pacifici, S. Rampino, A new bottom up approach to the CMMST VRE. VIRT&L-COMM.7.2015.8
133. A. Laganà, Research and innovative actions. Chemistry, molecular and material science and technologies virtual research environment (CMMST-VRE). VIRT&L-COMM.6.2014.1
134. C. Manuali, A. Laganà, Requirements of the chemistry, molecular and materials sciences and technologies community for evaluating the quality of a service. VIRT&L-COMM.5.2014.14
135. C. Manuali, A. Laganà, Trial user, resources and services quality evaluation for grid communities sustainability, Lecture Notes in Computer Science (2015), p. 324
136. J. Crawford, G.A. Parker, State-to-state three-atom time-dependent reactive scattering in hyperspherical coordinates. J. Chem. Phys. **138**, 054313 (2013)
137. X. Li, G.A. Parker, Theory of laser enhancement of ultra-cold reactions: The Fermion-Boson population transfer adiabatic passage of $^6Li + {}^6Li^7Li(Tr=1mK) \rightarrow {}^6Li_2 + {}^7Li(Tp=1mK)$. J. Chem. Phys. **128**, 184113 (2008)

138. X. Li, G.A. Parker, P. Brumer, I. Thanopulos, M. Shapiro, Theory of laser enhancement and suppression of cold reactions: the Fermion-Boson $^6Li + ^7Li_2 \rightarrow ^6Li^7Li + ^7Li$ radiative collision. J. Chem. Phys. **128**, 124314 (2008)
139. X. Li, G.A. Parker, P. Brumer, I. Thanopulos, M. Shapiro, Laser-catalyzed production of ultracold molecules: the $^6Li + ^6Li^7Li \rightarrow ^7Li + ^6Li_2$ reaction. Phys. Rev. Lett. **101**, 043003 (2008)
140. X. Li, D.A. Brue, G.A. Parker, New method for calculating bound states: the A1 states of Li_3 on the spin-aligned ($^4A''$) potential surface. J. Chem. Phys. **127**, 014108 (2007)
141. X. Li, D.A. Brue, G.A. Parker, S. Chang, General laser interaction theory in atom-diatom systems for both adiabatic and non-adiabatic cases. J. Phys. Chem. A **110**, 5504–5512 (2006)
142. R.T. Pack, E.A. Butcher, G.A. Parker, Accurate 3D quantum probabilities and collision lifetimes of the $H+O_2$ combustion reaction. J. Chem. Phys. **102**, 5998–6012 (1995)
143. G.A. Parker, A. Laganà, S. Crocchianti, R.T. Pack, A detailed 3D quantum study of the Li + HF reaction. J. Chem. Phys. **102**, 1238–1250 (1995)
144. B.J. Archer, G.A. Parker, R.T. Pack, Positron-hydrogen atom S-wave coupled channel scattering at low energies. Phys. Rev. Lett. **41**, 1303–1310 (1990)
145. J.V. Lill, G.A. Parker, J.C. Light, The discrete variable-finite basis approach to quantum scattering. J. Chem. Phys. **85**, 900–910 (1986)
146. M. Keil, G.A. Parker, Empirical potential for the $He + CO_2$ interaction: multi-property fitting in the infnite-order sudden approximation. J. Chem. Phys. **82**, 1947–1966 (1985)
147. M. Keil, G.A. Parker, A. Kuppermann, An empirical anisotropic intermolecular potential for $He + CO_2$. Chem. Phys. Lett. **59**, 443–448 (1978)
148. G.A. Parker, R.T. Pack, Rotationally and vibrationally inelastic scattering in the rotational IOS approximation. Ultrasimple of total (differential, integral, and transport) cross sections for non-spherical molecules. J. Chem. Phys. **68**, 1585–1601 (1978)
149. A. Schawlow, T.W. Hnsch, G.W. Series, The spectrum of atomic hydrogen. Sci. Am. **240**, 94–111 (1979)
150. C. Manuali, A. Laganà, GRIF: a new collaborative framework for a web service approach to grid empowered calculations. Future Gener. Comput. Syst. **27**(3), 315–318 (2011)
151. https://www.fosteropenscience.eu/
152. H. Goldstein, *Classical Mechanics* (Addison Wesley, Tokyo, 1964)
153. D.M. Hirst, *Potential Energy Surfaces: Molecular Structures and dynamics* (Taylor & Francis Inc., London, 1985)